Peterson, Brenda

Sightings: The grey whales' mysterious journey

Sightings

Also by Linda Hogan and Brenda Peterson (editors)

Intimate Nature: The Bond Between Women and Animals (with Deena Metzger)

The Sweet Breathing of Plants: Women Writing on the Green World

Also by Brenda Peterson

Build Me an Ark: A Life with Animals

Sister Stories: Taking the Journey Together

Pacific Northwest: Land of Light and Water (with Art Wolfe)

Singing to the Sound: Visions of Nature, Animals and Spirit

Nature and other Mothers: Reflections on the Feminine in Everyday Life

Becoming the Enemy (a novel)

River of Light (a novel)

Living by Water: True Stories of Nature and Spirit

Duck and Cover (a novel)

Also by Linda Hogan

Solar Storms (a novel)

The Woman Who Watches over the World: A Native Memoir

Mean Spirit

Seeing Through the Sun

Dwellings: A Spiritual History of the Living World

Power (a novel)

The Book of Medicines: Poems

Savings

Red Clay: Poems and Stories

Sightings

The Gray Whales' Mysterious Journey

BRENDA PETERSON
& LINDA HOGAN

WASHINGTON, D.C.

Published by the National Geographic Society
1145 17th Street, N.W., Washington, D.C. 20036-4688

Certain sections of this book have appeared in other forms in the *Seattle Times* and in the essay collection *Singing to the Sound: Visions of Nature, Animals and Spirit* (NewSage Press, 2000). *Fat* © Linda Hogan, from *The Book of Medicines* (Coffee House Press, 1993) reprinted by permission of the author.

Library of Congress Cataloging-in-Publication Data

Peterson, Brenda, 1950-
Sightings : the gray whale's mysterious journey / Brenda Peterson and Linda Hogan.
p. cm.
ISBN 0-7922-7989-1
1. Gray whale--Pacific Coast (North America) 2. Endangered species--Pacific Coast (North America) 3. Makah Indians--Hunting. I. Hogan, Linda. II. Title.

QL737.C425 P48 2001
599.5'2217743--dc21

00-054882

Interior Design by Melissa Farris
Printed in the U.S.A

For the journeys of the whales and the humans

and

to Larry Henderson, for his beautiful sightings

Fat BY LINDA HOGAN *xi*

INTRODUCTION *xiii*

GRAY WHALE MAPS 1

PART ONE
First Sight: Baja 5

PART TWO
Eye View: The Journey 55

PART THREE
Double Vision: Hunting the Whales 99

PART FOUR
Eye to Eye: Honoring the Whale 163

PART FIVE
A Cold Eye: The Far North 227

EPILOGUE 281

ACKNOWLEDGMENTS 285

"Whale watching takes on a whole new world of meaning when the whale is watching you."

Dr. Sylvia Earle

"The most beautiful thing we can experience is the mysterious."

Albert Einstein

Fat

This is the land
where whales were mountains
pulled in by small boats,
where fat was rendered
out of darkness
by the light of itself,

where what fell through the slaughtering decks
was taken in by land

until it became a hill made of fat
and blood, a town built on it.

The whale is the thick house of yesterday
in red waters.
It is the curve of another fortune,
a greasy smell and cloud
of dark smoke
that hides our faces.

At night
in the town
where hungers
are asleep,
we sleep
on a bed of secret fat.

A whale passes.
From dark strands of water,
it calls
its children by name.
Light, Smoke, Water, Land.

-LINDA HOGAN

Introduction

We come to the gray whales in all humility, with our senses open, our skills ready, and our imaginations engaged. Out there are whales, living by light and ancient brain, and our sightings must be insight as much as science, as if we are respectfully studying a still unknown culture. We come to the grays as ordinary people caught up in the mystery of creatures once considered unworthy of safekeeping.

In 1995 one of us, Brenda Peterson—a nature writer, journalist, and novelist—began writing about the plight of the Baja gray whale nursery and the Mitsubishi Corporation's plans to expand a saltworks on San Ignacio Lagoon in Mexico. In 1996, she was joined by the other author, Linda Hogan, a Chickasaw poet, novelist, and nature writer on the Native Working Group for reauthorization of the Endangered Species Act. Together we journeyed to Washington State's Makah Reservation, answering a call from Makah tribal elders who wanted to talk openly about their tribe's return to whale hunting. We did not know in 1996 that this historic meeting would begin a migration of our own to follow the grays. Over the past seven years, we have traveled the West Coast from Mexico to Alaska in search of the massive fellow mammals Jean-Michel Cousteau calls "breathing planets."

We have encountered not only the gray whales, but also the people whose spiritual traditions, daily lives, businesses, and sense of well-being depend upon these most-watched of all whales. "We really count on the gray whales migrating past our homes every fall and spring to reassure ourselves that everything is all right with our watery world," says one of the members of the Orca Network's Whale Sightings Network—an Internet community that reports daily sightings of grays, orcas, and other marine mammals.

This interdependence of whales and people is sometimes more impressive than the scientific facts of the gray whale migration. That is why our book is not a natural history narrative, but a multifaceted portrait of the human-whale bond. Writing this book in two voices—that of an American Indian woman and a nature writer—we hope to invite the reader to consider another culture, which happens to be nonhuman. The idea that animals engage in making and perpetuating their own cultures is not new. National Geographic Explorer-in-Residence Dr. Jane Goodall and other respected primatologists are engaging the world in a lively discussion about the cultures—and perhaps even legal rights—of the world's dwindling populations of great apes. Because we in the 21st century now face many difficult decisions about the futures of both human and animal cultures, this book is a passionate dialogue about both Native and animal rights.

These years of following the gray whales' massive 10,000- to 12,500-mile migration from their Baja, Mexico, birthing waters to their summer Arctic feeding grounds have left us with more questions than answers. Though this whale is now the most intensely studied of all marine mammals, it is also still the most mysterious. In *Sightings*, we share with the reader what we have learned about the gray whale, recent natural history discoveries, Native stories, and migration routes. We also tell you what we don't know, but can only fathom with our imaginations, because it is an animal so hard to define in words alone.

When we first began this book, gray whale migrations were a wildlife phenomenon that belonged mostly to the peoples of the Pacific Coast. Who could know that in the following years and into this 21st century, Eastern Pacific gray whales would take center stage in the world's international news? Today, their story is even more important, as Japan and Norway continue to lobby to lift the 1986 international ban on commercial whaling. Both countries are currently whaling. In 2001, Japan outraged many international conservation groups by expanding its scientific whaling to include among its total 50 Bryde's, 50 Sei, and 10 sperm whales—all species with declining populations.

In these years of our journeys, two significant events have shaped this dramatic environmental history. The first was the successful and highly controversial Makah whale hunt in May 1999, in which an indigenous nation returned to cultural whaling off the United States mainland for the first time in 70 years. Since that hunt, 14 tribes along the West Coast have signaled their intent to return to hunting the gray whale. Several non-indigenous peoples, such as Faroe Islanders and Japanese, Russian, and Norwegian traditional village whalers, have also claimed a cultural right to return to whaling.

Secondly, the spring migrations of 1999 and 2000 involved a record-high die-off of gray whales—278 known deaths in 1999 and 350 in 2000. Scientists noted that the whales seemed to be starving, their necropsies revealing lower-than-normal levels of fats. Added to this was a startling decline in birth rates, from a high of 1,430 in 1997 to an alarming low of 200 newborns in 2000 and 280 in 2001.

In 2002 scientists noted that gray whale calf births in Baja birthing lagoons were three times that of previous years. There has also been a marked drop in gray whale deaths in the spring migrations of 2001 and 2002. The great gray whale die-off is still a mystery. Was it because of climate conditions, such as El Niño? Chemical contamination in our seas? Scarce food sources? One recent theory about the die-off points to a possible shortage of crustaceans—tiny amphipods upon which the grays depend for food. Without enough food, the grays may have been too weak to make their mighty migrations.

Just as we document here the dramatic scientific and

political changes for the gray whales, we also tell the stories of how we each reckon with this massive fellow mammal: whether it's Linda's double vision of sighting the whale both as a Native person and a naturalist, or Brenda's struggle to report on the war often fought now between environmentalists and tribal councils.

Our dialogue on the whale hunt is one of the very rare public conversations between Native and environmentalist points of view. We take into account the traditional past when Native whale-hunting was a spiritual art of subsistence and reverence; we also bring the reader close to history being written at ordinary kitchen tables, where behind-the-scenes conversations between Natives and non-Natives are still going on. By looking beyond the headlines of conflict, we are doing what women have done for centuries: telling the story of the home front, the elders, and the next generations.

Yet the great whales are larger than this. They are beyond our rights, our treaties, and our histories. They exist beyond our own needs, in a world of land and sea that we share, but that we have never truly fathomed and never owned.

By placing our voices and visions together with the story of the gray whale, we hope to give the reader a diverse look at this mammal who still so mystifies scientists: Why do grays, of all the great whales, continue to initiate physical contact in the very birth waters in which 19th-century Yankee whaling Captain Charles Scammon slaughtered calves to draw in and attack mothers? Why do those of us who share the West Coast with this last healthy population of the once-great herds of gray whales know so little about their migration, their complex society, and their acoustics?

Here are stories of an interspecies bond that is both ancient and contemporary. It is a living history shared by land mammals and our mammal cousins of the sea. It is a kinship system in which a species comes back from extinction to offer us trust and an intimate relationship that still defies science. This trust calls us to consider issues deeper than just conservation, research, and education. It calls us to consider ethics and spirit when we think about a species besides our own.

We hope this book brings the gray whales as close to our readers as they are to our shores. Come with us on this journey with a whale who is older than human history, yet whose own history and future is now in our hands.

Linda Hogan, Colorado
Brenda Peterson, Washington

A Note on the Text: This is a book woven together in two voices. Linda Hogan's voice appears in italics and Brenda Peterson's in roman type.

Gray Whale Maps

I remember reading that long ago in European history, before navigation skills were learned, before a map of the world was known, men at sea often located themselves by the journeys of whales. Not yet having learned the stars as guideposts or the compass pointing north, it was through the whales that they found their human bearings. It was sensible, for certainly the map of the gray whale is ancient and reliable; it is brain deep, ocean deep.

In the minds of gray whales there is an inner geography, cartography unknown to us. It urges them along the electromagnetic directions of Earth. They read the polar constant from inside themselves. They know it by birth as well as by the placement of the sun. The mental maps of the gray whale are more powerful than anything we will know, as lost as we sometimes seem to be, even thinking we know our way. Along the Pacific coast, the whales travel a 12,000-mile round-trip toward the places where the sun is most bright. Their journey moves them between Baja and the rich summer waters of the Bering and Chukchi Seas. They watch the ever shifting world beneath as if they know the natural history of water, still mysterious to us, and far from constant. They have their own vision of the world, a waterway.

Because we inhabit a world that seems fixed, seldom do we think of a map as something that exists not on paper or chart. We don't often think of the figured movement of currents or tides; nor do we think of an ocean wave as cartography, a swell as a legend in a place liquid and wild. But in the seamless sea, the mind of the gray whale contains a map deep, fluid, and as shifting as the coastal currents. The whale has an elemental knowledge of its world, not only the shape of coastlines, point by point, but the radiance and direction of light.

Just as there are national and cultural maps where the world is seen as reflected by a country or a tribe, so there are whale maps. Their directions are told by the lengthening of days and the warming temperatures of the Pacific. They swim in their circle of life toward their food source. But they know the land as well. Points, bays, capes, and other marks are recognized, certain as cairns along the way. And when they rise out of water, above them they see not only the sun, the god they follow to the surface at death, but also the movement of the stars at night.

They read changes in the temperature of the waters, sometimes subtle, sometimes abrupt. They read weather in the air above, in a breeze, a strong wind. Theirs is a world we do not know or understand. Their pathway takes in a knowledge of the world beneath, the geologic and volcanic history of the Earth, the sea floor, the upwellings, fault lines, and the magnetic pull of the planet.

We have followed the travel of whales, looking for ones with recognizable patterns. "There is White Beak," we call out. Or the one with four spots lined up on its tail. We follow them because, maybe as the early Europeans thought, a whale is a map.

PART ONE

First Sight: Baja

Baja, Mexico, Spring 1998 and 1999

A Sheltered Place

Here in San Ignacio the shallow waters are warm, thick and silted with plankton, a buoyant broth. It is a sheltered place. This is why the gray whales come here to mate and give birth.

As we approach, from the airplane we see the water, seabirds, white and gray dunes blown in by winds. Islands appear at first—blinding white, a ray of sun igniting them in patches through the clouds. There are washes and arroyos around the water, then oceans and the shallow waters where the whales come to live between worlds of water and land as if they were a link between those elements. For a moment the clouds are all we see. Looking down, it is a world of clouds. Then sunlight.

The Last Lagoon

This is where it begins, the birthing and breeding of the great gray whales who make the longest migration of any mammal in the world. Each spring they start the treacherous trip north, new calves in tow—5,000 to 6,000 miles along the West Coast from shallow, warm lagoons in the Baja, Mexico, peninsula to summer feeding grounds in the Bering Sea, and back again in fall.

We are flying low over Baja, Mexico, in a 50-year-old DC-10, and the landscape is otherworldly: Surreal pink peaks rise above a pale desert on one side of us and a long, bright turquoise body of salt water lies bordered by a labyrinth of mangrove swamps on the other. San Ignacio is a shallow lagoon of intensely salty water that pokes into the Baja desert like a finger from the sea. It so remote, it has taken us almost all day to reach our destination: a peninsula that, from above, looks exactly like a whale's tail.

Linda Hogan and I are here to experience an interspecies phenomenon, *Las Ballenas Amistosos*, or the Friendly Whale Syndrome, that still mystifies scientists. Since the 1970s, when gray whales and their new calves first began approaching local fishermen in boats, people have been coming to San Ignacio in the hopes of an encounter with wild whales that is very much on the animals' own terms. For it is the gray whales who choose to initiate contact with people, not the other way around.

Though both Linda and I have enjoyed many wild encounters with other whales—from Alaskan orcas to Hawaiian humpbacks—neither of us has experienced whales watching and actually physically engaging with humans. A Makah tribal elder, Alberta Thompson, whose own visit to these Baja lagoons inspired her to protest her tribe's proposed whale hunt, first advised us to come to San Ignacio.

"Go see where the grays are born," she told us, holding up a photo of herself leaning far out of a boat, as a gray whale rises up under her hands. "You'll understand everything, and why I want to protect these whales for my grandchildren."

Our group of 18 whale-watchers is led by Jean-Michel Cousteau, who is here to film and document the challenges gray whales face in their struggle for survival. In an old bus painted with bright blue and purple whales, we drive down rutted dirt roads adorned with ancient shell middens to find ten weather-battered Baja Expeditions dome tents pitched on the beach. Mesmerized by the lagoon, which shimmers like a mirage, we see dozens of gray whales. Their misting spumes glow, backlit by the flaming desert sunset like brilliant blue-gray geysers each time a gray lets out a tall blow, followed by a shorter, smaller calf's blast.

As the fuchsia fireball of sun slowly dives deep into the lagoon, one small generator in our main tent sputters and whirrs. It is the only large, artificial light around. Flashlights pop and sway like fireflies. We see now more with our ears and suddenly hear an intense darkness, then everywhere the whooshing, solacing sound of many whales sighing.

Whales in the Desert

Something in the darkness of night calls out and it is large. It is swimming through the clouds of first day, rising from beneath. It is generous in what it gives and in its sheer power. It can seek revenge, this creature moving through water. Yet, in the night, if the wind is low, I hear its breath and the breath of its infant and wonder why, in this place and others—after countless years of whaling—they now come to us.

Then in the morning, just before the rose light of dawn is on the horizon, I'm awake, walking across the desert. Across the stretches of this desert, the plants appear dead, but are keenly alive. There are many lives here. I share my morning walk today with a coyote. Dark, ragged, we are little concerned about each other; she is only slightly wary on her part. Beside the path lined with shells, down to the water of the whales, are bobcat tracks.

Sitting in the sun, as I watch the spumes in the lagoon, the mist of mountains in the background, the breath from below rising behind the gray plants and sand, the magnificent body of a whale breaches. Up from abundance it rises, water falling from it, and then it enters water again. It is so large I can hardly take it in as it comes up from the rich waters. This mother who just breached has an infant alive in a suspension of earth, water, zooplankton, eelgrass, where the salt and brine offer buoyancy to the newly born.

Sometimes in my morning walks, I sit beside a whale vertebra, white and porous. There is also the skull of a sea turtle with its beak sharp. There are various other bone remnants, fragments, including one that looks like, perhaps, a blowhole with bone around it.

Coyotes, bobcat tracks, and the tracks of small mice, these are what I've seen in this place early in the day when I'm the only one up. Some days I go to the beach or the water's other edge. These and the whales in the dark lagoon are my companions. I hear them all night, breathing, blowing in the shallow, salted water. Then everyone is waking up.

Friendly Whales

Each night in the main tent, we talk whales, while in the nearby lagoon the grays breathe in syncopated blasts of air and mists. Above us, the Milky Way domes the dark sky and there are so many stars we bend under the luminous weight of them. Leaning forward in the circle of lights thrown by our lanterns, we listen to the history of the grays and their ancient birthing lagoons.

All whales trace their ancestry to ancient land mammals that returned to the sea about 50 million years ago. As they evolved in the water, losing their hind limbs and fur, developing a blowhole and protective blubber, they split into two suborders, the toothed whales and the baleen whales. Grays are baleen whales, straining their meals of tiny sea creatures through fringed plates of keratin inside their mouths.

There is fossil evidence that in pre-whaling days there

were four populations of gray whales in the world's oceans, including both sides of the Atlantic. But by the 17th or 18th century, the North Atlantic herd was extinct. The other Atlantic group was driven to extinction by the year 1750. Another group, the Western Pacific, or Korean, grays, was decimated by Korean and Japanese whalers. They are now endangered and down to under 300 individuals. The Eastern Pacific grays, the species breathing around us now in Baja, almost traveled the same deadly path. Twice they were driven almost to extinction: once by whalers in these very birthing lagoons in the 1850s, and again in the early 1900s by floating whaling factories. In 1946 the International Whaling Commission gave gray whales full protection, with the notable exception of indigenous subsistence hunting in Alaska and Siberia. In 1971 the U.S. placed the gray whale on the endangered species list. The Mexican government did its part as well, declaring the three Baja birthing lagoons at El Vizcaino Biosphere Reserve a whale refuge—the first such sanctuary in the world. In 1988 San Ignacio was chosen as a United Nations World Heritage site and Whale Sanctuary.

In 1994 the gray whale was taken off the endangered species list. This delisting occurred despite the objections of some marine mammal biologists. By 1998, its populations were said to have rebounded to an estimated 26,000. But in the spring of 2002 the grays had declined to 17,414.

As Canadian gray whale expert Jim Darling notes, "Gray whales hold the dubious distinction of being the only whale species with entire populations declared extinct." He goes on to note that "it is perhaps a sign of our desperation to succeed in conservation issues, as we declared gray whales recovered and watched them taken off the endangered species list [in 1994] to be hunted again. We know the fragility of the species. This is the last healthy population and we must look after it very, very carefully."

As we huddle against the night wind in our tent, we learn another history of the gray whale. This is local history, proudly told by Ranulfo, one of our boatmen guides.

In 1976 Ranulfo's father, Francisco Mayoral, was fishing in San Ignacio Lagoon as he had for many years, peacefully navigating the gray whale neighbors who kept their distance, floating near the fishing skiffs, surfacing and breathing their sonorous, bass blasts of air and mist.

The village fisherman had noticed that sometimes the mother-calf pairs approached his small boat as if curious. Several times, in their mammalian instinct for tactile stimulation, the whales even nuzzled, rubbed, or ran their long barnacled bodies underneath the boat, lifting it slightly with their knuckled backs. All mammals, it seems, crave touch, bonding. And these whales were no exception. What was different, Ranulfo's father noticed, was the mother whale's seeming trust of him with her newborn calf. Sometimes she'd even let her baby come right up to the fishing boat, as if seeking to be stroked, while she carefully cruised alongside.

Then it happened, and at the whale's initiation. "First a curious calf, then a mother..." our naturalist Chris Peterson took up. "Then mother-calf couples, then some juveniles, even a bull or two—slowly all these whales began to seek out not only the boats, but the humans in them. And that was the beginning of this mystery we still call *Las Ballenas Amistosos*, and scientists call the Friendly Whale Syndrome."

In 1977 California gray whale experts Steven L. Swartz and Mary Lou Jones first documented this phenomenon for NATIONAL GEOGRAPHIC magazine. Encountering a friendly two-year-old juvenile whale they named Amazing Grace, the researchers told the story of being "adopted" by Amazing Grace as "her personal toy." Curious and playful, Grace engaged with Swartz and Jones. "She would roll under the boat, turn belly up with her flippers, sticking three to four feet out of the water on either side of the craft, then lift us clear off the surface of the lagoon, perched high and dry on her chest between her massive flippers."

After such gymnastics, Grace would seek contact from boats and people. "We would oblige her with a vigorous massage along her back, head, and ribs, while she opened her mouth to display huge fringed curtains of creamy white baleen plates." Sometimes friendlies even knocked people out of their boats in their enthusiasm to make contact. But Grace's sociability with humans was not typical of all the grays in San Ignacio, according to Swartz and Jones. "Friendly Whales find you," the researchers noted, "you don't find them."

Soon word got out about these Friendly Whales in Baja. Over the next 30 years, many cetacean lovers, scientists, and adventure travelers made the long pilgrimage to encounter these grays, who seemed to be defying the logic of self-preservation. Despite their near extinction at our hands, they were again seeking us out.

A Whaler and a Naturalist

The whales in Baja's bays and lagoons were once so thick that the whaleboats couldn't get through the waters. At San Ignacio the boats only had to wait at the mouth of the lagoon to get the whales as they came and went. The whalers killed grays in unprecedented numbers. By the 1850s, there were over 50 ships in the small lagoons, blocking the entryways. Less than 30 years later, the boats and the whales were gone.

Captain Charles Melville Scammon, whaling captain and naturalist, found the gray whales' breeding grounds at Ojo de Liebre in 1858. Scammon reduced the population to near extinction, but also wrote about the whales in a book entitled The Marine Mammals of the Northwestern Coast of North America; Together With an Account of the American Whale Fishery. *He not only wrote about the close bond between mother and infant, but he also killed the animals and described their bodies, the embryo of infants, the feeding of milk, the love of the mothers who would fight to protect their infants when his men lanced and killed the babies.*

The whales fought back. Scammon wrote about the upsetting and staving of boats. The mothers would overturn them. Whalers called them devil-fish, as if the whales were evil for fighting for their lives.

Scammon drew the gray whale fetus, in all its beauty, in his book. The most marvelous thing is that the fetus has vestigial teeth in the upper jaws that disappear as it grows. These teeth become baleen, the combs through which the food at the bottom of the water is strained. Strangely, this unlikely book is considered one of the best books about the most primal, most mysterious creature on Earth, in its waters.

Trust

Gathered in our tent as the warm wind sweeps across the birthing lagoons, we marvel at this history of human and whale contact. "Why do they like us, I wonder?" asks a woman who spends her early retirement as a self-taught naturalist on whale-watching expeditions in southern California.

Glenn, our lanky camp director, responds, "Here we've created a balance between human and nature that encourages the whales' native curiosity. In other places along the whales' migration there is no human invitation for interaction—just cargo traffic, motorboats. Whales are run over, ignored, or hunted. But here in the lagoon we've created an environment of trust."

"Should we teach them to trust us?" a man asks. He has gone whale watching for over a decade and never experienced anything like this lagoon. "What if they learn to trust humans here and then up north, the whales come up to boats just like ours? And instead of our hands and touch, they meet hunters?"

No one answers. Many of the Alaskan Inuit and Russian Native peoples still subsist on whale meat. For decades, whalers from the former Soviet Union killed over 100 gray whales a year in the Arctic and Chukchi Seas, using the whale meat to feed foxes in fur farms that have recently been abandoned. In 1997 the Russians were granted a five-year quota of 620 gray whales. Many of us in the tent dread the imminent whale hunt by the Makah tribe.

After a long silence, Glenn quietly gives us a gray whale education. Grays are known for having "knuckles"—not dorsal fins, but a ridge along the back. The female gray whale is slightly larger than the male, measuring up to 46 feet, and both sexes weigh 30 to 40 tons. At birth the calves average 14 to 16 feet and weigh 1,500 to 2,000 pounds. The milk of the mother is rich, 53 percent fat, and so the infants gain 2.5 pounds an hour, 60 or 70 pounds a day, doubling their weight by the time they leave on their migration. The calf will not be weaned until six or seven months, and while nursing young in Baja lagoons, these gray mothers lose much of their own stored body fat and blubber reserves.

There is still much to be learned about the feeding patterns of the gray whales. How can such gigantic mammals survive on the tiniest crustaceans, such as beach fleas, sand hoppers, and other amphipods buried in the ocean muck? In their gigantic, three-chambered stomachs, scientists have discovered hundreds of pounds of these amphipods.

Grays are unique among whales because they are bottom-feeders. Using their great maw of a mouth as a shovel, they suck up the sea floor's fertile sediment, squeezing it through their delicate baleen filters with huge tongues. We can tell whether each whale is a right- or left-sided feeder by which baleen plate is shorter. Grays will also have deeper scars on the side of their mouth they use for feeding.

When and where exactly they feed is still somewhat in debate among marine mammal biologists. Many believe that grays eat only in Arctic seas. Still, in San Ignacio Lagoon, whales have been observed trailing eelgrass and seeming to forage through it for fish and perhaps grubs. But the shallow sea bottom here is not fertile and teeming with amphipods, as are the main feeding waters of the Arctic. Scientists also believe that in their urgency to reach Baja breeding and birthing lagoons, the gray whales rarely eat on their southern migration. On northern migrations whales often pause to feed on seasonal ghost shrimp beds, especially around Whidbey and Vancouver Islands. Some grays will stay near those islands to become summer residents, while others migrate far into the fertile Bering and Chukchi Seas.

All these impressive facts, feeding patterns, and measurements do not prepare us for the aerial photos our expedition leader shows us of whales *double* the size of our 18-foot skiffs. There is murmuring in the tent before we retire to our sleeping bags. How do we really prepare to meet such great and ancient creatures, except by first closing our eyes and listening to them breathe—floating through our dreams?

Next morning, our marine biologist, José Sanchez, schools us in whale etiquette as we excitedly climb aboard our skiffs and ease into the shallow lagoon. "Some of these whales show harpoon scars in their skin. Yet they still seek contact with us, especially in this lagoon," José says. "Imagine a 40-ton tractor-trailer truck and you're in a VW bug alongside—that's the ratio of whale to our boats. With one flick of a tail fluke or nod of a huge head, we'd be tossed overboard. We are completely at their mercy."

San Ignacio is the last pristine and protected gray whale nursery. The gray whales can come here as they have for hundreds of thousands of years. Yet at the time of our visits in 1998 and 1999, the lagoon was threatened by a controversial proposed expansion of a Mexican saltworks, jointly owned by the government and a private corporation.

But thanks to a massive grassroots environmental movement—including action by Mexico's Grupo De Los Cien, the International Fund for Animal Welfare (IFAW), the Natural Resources Defense Council (NRDC), and a campaign involving over a million citizen letters—San Ignacio Lagoon would later be saved. In June 2000 the global tidal wave of concern resulted in a cessation of the saltworks expansion on this vital lagoon.

Perhaps this act of environmental responsibility is the first sign that the 21st century might help to heal the scars of our history toward other species. It also shows something else: It will not be just the governments, the scientists, or the corporations that save the Earth and other species; it will be, first and foremost, the people.

There is still much work to be done for the gray whale to flourish into other centuries. So recently removed from the endangered species list, the grays now face what Jean-Michel Cousteau calls "the obstacle course" of their migration journey. One-third of these calves will not survive their first year and a journey filled with peril: Boat propellers slash their young bodies; gill nets and drift nets entangle their flukes and drown them. Pollution from our industrial heavy metals and PCBs poison these baleen bottom-feeders, and recent U.S. Navy coastal sonar tests disorient the whales. Age-old hazards include transient pods of killer whales that sink their teeth into the calves' throats, even as the mothers struggle to lift their newborns on their bellies and away from the predators.

Here in San Ignacio Lagoon, the Mexican government wisely limits ecotourism to 23 permits per year issued to Mexican locals. Only 12 boats are allowed in the outer lagoons at any one time. People such as José Sanchez, Ranulfo Mayoral, and Martin Aguilar Arce from the nearby small village enjoy sharing their lifelong expertise on the native flora and fauna—and, of course, the whales.

In the main tent, Martin proudly shows us photos of himself leaning out of the boat as a gray whale calf spy hops almost into his open arms. "I love the whales, *las ballenas*," he says somewhat shyly in a soft voice. He is learning English and has just graduated from a prestigious course in marine biology.

"So far back, the whales and the people go together," he murmurs. He shows us 1,000-year-old Perico, Guaycura, and Cochimi petroglyphs of whales. Pictures of sloping humps, spiny knuckles and wide whale faces, these petroglyphs are scrawled alongside human figures.

Old Light

I heard a tale that the sun god was said to live in a whale and that's why light seems to come out of its eyes and why rainbows form in the mist of the gray whale's breath. It's true, light comes, but it is an old light, seeing. They look. The way bathers lift themselves from the sea; they are shining, water falling from them as they rise, exhale, inhale and return below the surface of the water.

The gray whales themselves are an intelligence we haven't yet grasped, understood. Life-covered with several hundred pounds of barnacles, the whales are small-eyed. If we could see them over the brief timescale of the planet, they would look like shape-shifters. If we thought of time by something other than our own notions, million-fold years ago, before our own knowledge of evolution, we'd see these whales as they walked on land. Even now, existing within their immense bones, there is body evidence, a hand, human;

vestigial hipbones; and remnants of legs they no longer need. They lived on land when we lived in water. When we look at them we are looking at the past, and in another being perhaps we see our own future.

And when they come up again for air and the water falls from them back into the seawater, it is a shine of beauty in a world of desert, dunes in the background. As their hind legs have disappeared into them, out of our sight, and their hands have become hidden, the compass set into their brains, you'd think you might be able to be one with them, as with a cousin, but it's a cousin lost in time. It makes us so small in the firmament. It makes us remember something we can't quite name, only feel.

Our naturalist tells us, "I think of the grays as amphibious, because they live so close to us in the shallow waters off our shores. They really are creatures of both air and land, more than any of the other great whales."

Looking at these whales in the shallow water, thinking of the great turn of evolution and change on earth, we wonder if one day we will return to the seawater, along with these long-enduring, longest-living mammals on Earth.

They are a mystery to us in many ways, as we must be to them when they surface to look at us from their other element. Clear-eyed, they take our measure. This is the only thing it can be called. Measure. Looking at one, trying to find a physical description, the eye so far back the whale seems alien to the human notion of anatomy. It is a

mammal, yet it has a beak, whiskers, sea lice, and barnacles. Some might think it is an animal not to be loved for its beauty. Some think of it as their water sister or brother. By some, it is loved for its mystery.

And the whales too have a consideration. What is a human, they seem to wonder, and it's what the whale also prompts us to consider, as if we are reflected here. What is a human, the mammal that evolved out of the sea, the same sea they entered. Wondering this, deep in our cells, is perhaps why we seek them, why humans want to touch them. Perhaps that is why they come to us, we come to each other, drawn together, like relatives lost somewhere along the scale of transformations. In Earth time, it is only the way a tide comes in and returns, carrying us with it.

Contact

Martin has his own soft whistle for calling the whales as we skim along the lagoon in our skiffs through the no-whale-watch zone toward the opening of the lagoon. Some of us shiver in the early morning air. Though we are in a desert, this inland sea is blessed with a constant cool breeze so that in the shade or on the water, one has to gear up with rain slickers and even galoshes. Nights, we sleep in knit caps and wear gloves in our sleeping bags.

Mornings, some of us can hardly recognize each other from the night before. But what is recognizable, even palpable, is our excitement and anticipation as all of us await sight of our first gray whales, close up.

The two children on our boats soon prove themselves to be what someone calls "whale magnets." The whales are most curious about these little humans leaning way out of the boat and laughing, their hands splashing the water as adults hang onto the children's life preservers to keep them from falling in.

Suddenly a mother and calf approach the children's skiff in slow, graceful glides. The whales are huge, like giants sliding under a toy dinghy. Startled shouts, as the mother whale gently lifts their entire boat up over her knobby back, then lets it settle into its rather comic, cork-like bobbing next to her massive body. The children howl with laughter and run to the opposite side of the boat in time to see the calf lift straight up out of the water, only a foot from their outstretched hands, its curious eye taking everyone in.

"Spy hop! That's a spy hop!" the 12-year-old boy calls out, thrilled, as a calf surfaces, snout-first until its eye is above the water, then slowly pirouettes as if to peer around at the people.

Then the calf simply sinks back into the waves and rides the mother's wake away from our boat.

Planet of Water and Whales

In the universe called ocean, the gray whale is like a planet, says Jean-Michel Cousteau. "What are they?" he asks, coming here to Baja to study and photograph them. No one knows, but those who are interested are dedicated to understanding.

Myself, I think smaller, perhaps a continent. Certainly it is covered with life. It may be that in the planet of water, they are ensconced with creature life dwelling on them. There are barnacles and sea lice on their flesh, with journeys, labors, and survivals of their own. No one knows if the barnacles are symbiotic, beneficial companions, or painful parasites. We use these animal topographies; at least we try to use them as identification marks, along with visible scars from orca bites and tooth marks, as we follow them on their northern migration.

When the whales lower into the rich salt waters of the lagoon in San Ignacio and other waters along their migration, the barnacles *open and the animal inside emerges. Then the whales surface and the barnacles close, the sea lice move from place to place. When the baby scrapes the mother's barnacles, the calf then begins its life covered with other lives.*

At times the gray whale leaves water briefly and goes to lie in the sand to feed on invertebrates or rest. Sometimes the whales seem to almost live on the beach. Grays enjoy the warmth of sun and sand and rub themselves into the sand as if scratching their backs. Sometimes, too, they worm their way through the wet sand and mud, leaving a body trough.

To those of us who have come here, their length and girth are beautiful as they bound through the water. We are here to see them raise their shining infants from the windswept plankton, fertile waters, the place with the desert all around, a place of great contrasts, where the water temperature varies from 50 to 90 degrees, tidal currents are stronger than most rivers, where it is cold in the morning and night, hot by day.

The tide fluctuations here are dramatic; and it's a new moon. I am always amazed by the effects of moon on water, including the blood in our own bodies. But here it is visible. Inside us, only felt. In the tent, we have constant wind coming in from the water, crossing the flat, desert land, inseminating, as if giving life. Sometimes the wind and sand fill the night, even the tent. We try to stop it from entering by using clothespins on the tent's vents, although by day in the heat we put cardboard in the flaps to keep them open. At night when the world is dead silent, you hear them, whales breathing as they rise to another element. You have a sense of awe.

Later, finally in our skiffs, we see the whales, one belly-up. Mud flows from their mouths, one mother bringing up the infant on her back, giving us the eye. Perhaps the nursing mother leaves a trail of milk. She watches us, too. The old, old eye. Inupiat Eskimos have discovered stone harpoon points in five bowhead whales. These, as well as studies of the whales' amino acids, indicate that bowheads live from 90 to over 200 years. Gray whales are believed to live about 70 years, but clearly whales may have a longer life span than we've imagined.

At times I almost think I hear them underwater, low sounds, deeper than most. A rumble, nearly subsonic. Most people can't hear them, but some can catch their deep sounds. I recall reading about the 72-year-old man in Alaska who could hear them. Too old to hunt, his job then was to tell the whaling crew where the whale would surface.

When we see whales spy-hopping, we look for reasons. Some say the whales are troubling the ocean floor, stirring it, turning the food. We'd like to think they are looking at us, because as humans, we want to think everything has to do with us. Yet, as a member of this species, not even understanding my own self, I do think this; that they look out from these mostly shallow waters at us and the other lives above the surface of their world. The head comes through, entering another element, and you see it take everything in.

Breathing spume with rainbow in it from light, I think of our need *for them, not just this experience, but our reach for relationship and love, the only word I can use to describe the people I watch as they touch the whales.*

The gentleness of their great lives toward us is as much a mystery as the human need to touch them. I have two minds. I care for them. I want to see them, and yet I think it can't be good for them, this offering to us. Where they bring their babies as if to say, Protect this holy child, this life of my life.

Looking at the baby whales, I want to tell the infant to stay away from our kind at the same time I also want to be with the calf. This is the dilemma. That I want to care, knowing that others want to kill them. Then I think we should be chasing them away. Instead, the humans here want to touch them. I watch the spouting, the blow.

I see the skin of the mother and calf, so beautiful and mottled, like a ground of lichens. I see something I can't name in the eyes, the eyes watching us. The calves are born tail first. I can see, in my mind's eye, the newborn tail emerging into the water. Their first beautiful entrance into the water world.

Body Language

"Coming up!" a woman in our boat shouts as all six of us shift to one side of the skiff. Delightedly we slap the water to call the whales, whose underwater life is all about listening. Some scientists believe that the original impulse for the whales to come to boats was the gentle chugging of the engines, a bass sound similar to the low-frequency grunts and Chinese-gong-like vocalizations gray whales exchange.

Other researchers believe the whales use our wooden skiffs to scrape barnacles and sea lice off their sensitive skin; still other theories hold that the gray whales in their birthing lagoons have been protected for enough generations to build up a bond of trust between humans and whales, so that their seeking contact is actually a form of kinship or communication. But what are they telling us?

We, humans and whales, share body language. As the great behemoth mother whale swims toward us, my moment's fear gives way to awe as a long, slick head lifts up before me. Water streams down the mother's barnacled face; I see only her huge eye inches from my own.

Unblinking, calm and considering, this otherworldly eye can scan underwater mountains and shallow shoals; this eye has seen humans with harpoons and rifles hunting in Arctic waters; this eye is wary of hunting orcas, as well. This eye has a close regard for the newborn calf always nearby. And this eye also tenderly takes in humans who lean now half out of the boat in an offering of outstretched hands.

Resting my chin on the boat, I exchange a long look with the mother whale. She and her newborn calf seem to be resting beneath our boat, and she is rocking us slightly with her belly as if we are humans in her tiny cradle. Then the mother dives beneath our boat, her huge body a blur of white barnacles and dark gray leviathan. But behind her is the calf whose head surfaces with a blast of breath from double blowholes.

The baby's breath mists over us, smelling fresh and salty. On the northern migration, where whales begin feeding again, their breath is full of brine and the pungent swamp-scent of krill. But here where the birthing begins and the mothers nurse, the calf's breath is strangely sweet and surprisingly warm.

Coming near us, the calf rolls to take a closer, more curious look. This eye is smaller, more animated than calm, and

I wonder if perhaps we are the first humans this calf has encountered in the first month of life. Many of us feel a profound protectiveness and a sense of responsibility when the calf decides to trust us and offer a glossy snout for our touch.

The calf seeks our palms as the mother eases alongside, baby-sitting, but not begrudging her newborn time to play. Are we bobbing toys to this baby? I wonder. Are we playthings? If so, the baby calf is more good-natured, coordinated, and careful than most human children as he gracefully raises his sleek snout to the woman next to me.

"Oh, you're a beautiful baby," the woman croons as if singing to her grandchild. "Just look at you," she marvels, rubbing the white patch of barnacles adorning the calf's mouth.

What is most striking about this baby gray is the albino beak that gleams in the sunlight. "Let's call him White Beak," Linda suggests, reaching far out of the boat to scratch the small whale's bright snout.

We are all looking and touching. We cannot help but touch, as White Beak insists his long snout up into our waiting hands, then pirouettes in the water, inhales a quick, baby's breath, and dives back down.

As we float alongside the mother whale, her calf once again rises to seek us out. He lifts his lovely, long gray face up to be stroked and one of the mothers breathes, "Safe journey, little one, safe journey."

On the last night of our stay in San Ignacio with The Friendlies, the wind suddenly stops. Startling stillness awakens me from my tent to crawl outside and listen. A quarter moon hangs in the sky. I need no lantern to make my way along bright seashell-lined paths to the shore. Before I reach the beach I hear the long, sonorous sighs rising from dark waters—whales breathing like a lullaby. Whales blow, the surf laps, somewhere a camper snores, while above the heavens are alive with a multitude of stars, constellations so plentiful I don't recognize this ancient night sky.

Listening to mother and calf and humans resting, perhaps dreaming together, I look up at a universe our human lights usually limit. Here in Baja there are all the stars we never see, like the many invitations between ourselves and other creatures that we never answer.

What might happen if for century upon century, we keep this place whole? What world might our children witness after generations of an unbroken bond with these other mammals? As I sit on that moonlit beach, I know this is not only a whale nursery—it is also place of birth for humans.

On Earth Before Us

Whales today, all around. In this afternoon's group, we are surrounded by four mothers and their calves. Looking into our little skiff, touching us as we touch their skin, whiskers. In the distance a baby opens its mouth and the baleen inside is almost orange and so large you wonder how they can hold those plates, sieves, in the mouth.

In this lagoon, as at Baja's Magdalena Bay, Guerrero Negro, and Scammon Lagoon, once called Ojo de Liebre, whalers killed huge numbers of whales in the middle of the 19th century.

Conclusions from logbooks and sea journals, wrote Captain Charles B. Hawkes in a 1924 book entitled Whaling, *reveal "that the old whaling vessels had more than their arithmetical proportion of madmen." Here was destruction so great, it is a miracle the whales ever returned and are here before us. Madmen, when it comes to whaling and other animal kills, are never tried, though the judgment goes down in history. Because of this history, after a loss so great, so large, we are fortunate to be in the presence of these largest among creations on Earth.*

In our time here in the lagoon, the water, even with its currents, seems away from the breathing, the pulse, of the ocean, the rest of the planet. We see no swimming schools of fish. Outside is the spinning of earth, but for a while it is forgotten. And then there is no need to dream; the world is full enough.

Many people feel themselves called to the Baja birthing lagoons to witness and meet the gray whales in their own element. One of these, activist and naturalist Will Anderson, first journeyed from Homer, Alaska, to Baja in the late 1970s.

"The gray whales had captured my imagination," he said, "but I relied mostly on films and books. I really wanted to experience these whales firsthand."

So he exchanged his Arctic parka for a sun hat and journeyed down to Magdalena Bay, Baja. In the late 1970s, scientists were just beginning to document the Friendly Whale Syndrome.

"My first year down in Baja," Anderson said, "I just had a cheap army-navy tent and a one-person inner tube of a 'boat.' I'd float out in the currents near Puerto Lopez Mateos, a fishing village that at that time was at the end of a long dirt road. This waterway is just south of the entrance of Boca de la Soledad, north of Magdalena Bay." Here, at last, he encountered the gray whales who had so captured his imagination. And for the next three years, he spent all the time he could afford in Baja with the whales.

"In the quiet desert air, far away from people and town, all I could hear was the breathing of whales—24 hours a day," Anderson recalled. "I went to sleep and woke up to this sound of whales breathing.

"I had a rule," Anderson continued in a bemused voice, "that I would not touch a whale. But so many times, gray whales would come to just look at me from five feet away. A whole parade of them! I had become the observed. I started feeling the rhythm and the pulse of their lives—and this changed my own life."

Once Anderson was even sunk by a gray whale. "I'd move my camp to follow the gradual movement of the calves towards the open ocean as they grew stronger," he remembered. "One day I was listening to whale vocalizations beneath the water when suddenly a gray surfaced right under my kayak. At night they would usually avoid my kayak because I light up the hull like a jack-o'-lantern. When this whale surfaced, I was raised high up in my boat in broad daylight on this big whale's back! I just froze. And then the whale very slowly just sank back into the water. I put my kayak paddle in the water to move. But the next thing I knew—I was airborne, propelled upward by the whale tail! Smack! My kayak landed upside down. I fell straight out in midair and was pulled underwater. At last I surfaced."

Anderson was shaken but all the more fascinated by the gray whales of Baja. They have become a lifelong commitment. "Baja gave me a perspective that was more like that of the whale's than of people. It's so essential that we human beings develop empathy. Once a person crosses that threshold, you start looking at all life differently. You have to wonder, What are the animals feeling? What is *their* experience at the hand of humans?"

The Greatness of the Whales Toward Us

I remember here the human need to touch a great life. Here where they were brought low by whalers, here where once there were none. So I think of this desire to touch one. Everyone felt it, as if it were a part of human emotion or need, a movement toward wholeness and body integrity. As photographer Steven Meyers says, the whole body sees. The gentleness of the whales toward us is again remarkable. I can't know why they seek us out.

Perhaps their memories are short. Perhaps they are asking for help. Perhaps they are curious about us, the mysterious humans, and forgive or forget us and display for us their babies. There is such largesse of creation and sometimes it includes the generosity of other species. Surely this is true here. Even now they teach us about ourselves. We are needy. We want something. We ask a question here without language, ask it of the whale. We seek an answer to the unknown question that comes from a secret place inside. Or more likely here a secret place outside ourselves—a whale.

They are born shining and smooth, free of the barnacles that will soon live on them. In Baja lagoons, seagulls descend to eat the afterbirth.

Desert lagoons, sunlit habitat, the magic of how light and water becomes organic matter. In large numbers, unseen by the naked eye, are living cells. This place is full with life.

Amazing Grace

On our second trip to Baja's San Ignacio Lagoon in March 1999, we arrive to an unsettling sight—the carcass of a female gray whale decaying on the beach near our base camp. All spring, reports of 50 dead gray whales off Baja's peninsula—the victims of alleged industrial pollution—have added to the alarm raised when earlier this year, 18 dead whales were discovered in the Ojo de Liebre and the Guerrero Negro lagoons. In those places, saltworks were already accused of criminal negligence in killing endangered sea turtles and other wildlife with a spill of toxic brine.

The dead whale at our expedition base camp is a daily reminder of the dangers these grays face. At the end of our week of enthralling close encounters with gray whale mothers and calves, someone suggests that we give thanks for our interspecies bond and commemorate this one great whale's passing with a beach ceremony.

As the cormorants cry overhead and the mother-calf pairs loll right offshore, their breathing like a sighing chorus to our small memorial service, someone says, "This whale has made the migration full circle—returning in death to the lagoon where she was born. We are so grateful for the gray whales who share our shores and journey alongside our daily lives like close relations."

A grandmother adds, "We pray that if she had a calf it did not suffer long or was adopted by another."

Another man echoes, "I have never felt such generosity of spirit from another creature as I've been given here by these gray whales. And I will do whatever I can to protect this place of peace and birthing for the next generations."

"If we can just keep this San Ignacio whale nursery safe," someone murmurs softly, "there will be more hope for us all, all creatures."

Then in memory of the first Friendly Whale researchers had documented over 22 years before, we all sing "Amazing Grace," with the great, beached body of the gray whale behind us on the grooved sand.

Gazing at the blue lagoon where the breath of many whales blasts brief clouds on the horizon, I remember that the whole world is also a nursery—a birthplace for what will come next. As we sing, the sun eases away from its morning mist and illumines our small circle—our memorial for a great life lived so near our own. Nearby, the whale's body is like the silhouette of a sloping, humble mountain. And we sing louder, for all the birds and all the other living whales to hear.

PART TWO

Eye View: The Journey

San Diego, California, to Whidbey Island, Washington
Summer 1998 to Spring 2002

Interspecies Friendship

Every day we hear of acts of heroism and compassion by humans, stories of friends along the way who celebrate the gray whales' seasonal journeys with ceremony, festivals, rescues, and even prayers for safe passage. From the coastal Stranding Networks, to the whale census takers, to the Internet Sightings Network that tracks whale migrations, to Native hunters who sometimes rescue whales trapped in Arctic ice—these people weave their own lives into those of the mammal cousins who share their coast. These friends of the gray whale were never more important than during the die-off of grays during the 1998 and 1999 migrations. And even now, when there seems to be a small rebounding in births and populations, scientists need all the help they can get from amateur naturalists to unravel the mysteries surrounding this most watched of all whales. Here are just a few stories of this interspecies friendship.

"The Best Part of My Life"

It was a devoted naturalist and stranding volunteer, Terry Rogaczewski, who helped rescue a newborn gray whale and in so doing greatly advanced our knowledge of the hardships facing infants on their first migration. At birth, infant gray whales appear almost otherworldly with their fine, silken baby whiskers, their smooth silver skin not yet covered with barnacles, their eyes curious yet ancient, their bodies still wrinkled from the womb. To encounter such a newborn is a wonder few of us ever experience.

When a seven-day-old baby gray whale—emaciated, hypoglycemic, and disoriented—stranded herself on the beach off Marina del Rey, California, in January 1997, volunteers and lifeguards valiantly pushed her back into the sea to search for her mother or other southward migrating whales. Twenty-seven-year-old Terry Rogaczewski, who was one of those volunteers, kept an all-night vigil by the water. By morning's light, he discovered that the newborn whale was still alive, but very weak, floating in the marine channel. Because of the Marine Mammal Protection Act, which prohibits harassment or removal of any marine mammals unless given federal permission, Rogaczewski and others in their small boats waited until given the authority to act.

And then act they did. Rogaczewski and several lifeguards plunged into the chilly channel and swam alongside the baby whale. Though still only days old, she was over 1,600 pounds and almost 14 feet long. Even for Rogaczewski, who spends most of his time rescuing stranded seals, dolphins, sea lions, and whales, the rescue of this baby whale was momentous.

"To be right there and to be swimming in the water with her and to be part of her rescue," Rogaczewski said, "that was the best part of my life.... When we took her off the beach, she was so skinny, you could see the skull and the ribs...."

Hoisted onto the flatbed of a truck by a small army of strong volunteers from 15 different wildlife rescue organizations, the baby whale, with her umbilical cord still attached, survived the 112-mile trip to SeaWorld in San Diego. By the time she arrived, she was comatose, dehydrated, and severely malnourished. In the aquatic intensive care unit at SeaWorld, the baby whale was nursed round-the-clock. She was given the name "J.J.," after the late Judy Jones, a registered nurse and the longtime director of operations for the Laguna Beach-based Friends of Sea Lions Marine Mammal Center.

For J.J., her good fortune in being rescued by Rogaczewski and other volunteers from the Marine and Mountain Wildlife

Rescue team in Malibu gave her a chance to learn how to play, but in vastly different circumstances from the Pacific. In her 40-by-40-foot tank, J.J. was tube-fed antibiotics and nutrient-rich milky formula to simulate her mother's milk. Blood tests showed no signs of infection so far. Soon she was showing a sucking reflex. Every three to four hours, J.J. was fed two gallons of her formula, nursing from a foot-long nipple.

In the first weeks of J.J.'s rescue, SeaWorld received over 11,000 calls from people concerned about the baby whale's survival. The plan for the calf was to help her recover from the trauma of losing her mother within days of birth. The second step was to actually return her to the wild. This was a huge undertaking that had been successfully accomplished only once before at this park with a whale named GiGi (short for "Gray Girl").

Other infant gray whales are not so lucky as J.J. and GiGi. Within a week of J.J.'s rescue, another baby gray got entangled in a kelp bed off the San Diego coast. While rescuers awaited federal permission to free the baby gray whale, the little calf grew lethargic. Even once clear of the kelp wrapped around her tail, the calf did not swim away. Without any sign of a mother or other whales around, the calf listlessly floated awhile and then sank to a watery death. Calves such as J.J. are the rare success stories, small miracles of interspecies bonding.

By late January, J.J. weighed 1,840 pounds—still several hundred pounds underweight, but a sure sign that she was holding her own. To the delight of her devoted caretakers, the calf was swimming by herself and soon was demanding to be fed.

Over the next year, as J.J. grew up, she captured the

imaginations not only of the children who visited SeaWorld to press their faces up against the glass wall separating them from this massive mammal cousin, but also of the world. "J.J. Updates" were a feature of news reports as far away as Tokyo. When at eight months she was weaned off a formula of pureed fish and vitamins, J.J. began consuming 475 pounds of fish a day—from herring, squid, and krill to sardines, small shrimp, and capelin. In the wild, of course, she would have fed on the tiny benthic amphipods and crustaceans in the ocean bottoms. But her daily diet of 150,000 calories nourished her well.

When her first rescuer, Terry Rogaczewski, visited J.J., he noted "she was just rolling in blubber." Citing his personal ethic of animal rescue, he concluded, "My feeling is that we've done so much to them over the past, we kind of owe it to them."

A little over one year after Rogaczewski first found the emaciated gray whale, J.J. had grown to a splendid 30 feet long (a full 18 feet longer than when she first beached) and weighed in at 18,000 pounds. At 14 months, J.J. was the largest marine mammal ever held in captivity. Many worried about her scheduled March 31, 1998, release date. Would she be strong enough to feed and defend herself from the myriad threats to her wild survival? What would happen if she encountered orcas or other grays? Would J.J. recognize the latter as her own kind and the orcas as her predators?

Scientists debated many of the mysteries about gray whales during J.J.'s 14-month captivity. How much of a whale's knowledge of predators is inbred, and how much learned from the mother? What triggers its migration instincts? What is the purpose of the gray whale's acoustic rumblings? J.J.'s time on land also gave researchers invaluable gifts: They used her blood to develop an antibody serum to help other stranded or ailing whales along the coast.

Because J.J. would be restored to the sea, a decision was made to keep her "as wild as possible," said SeaWorld's senior animal care specialist, Kevin Robinson. "We don't want her to be looking around for humans to feed her or pet her." It was a good sign that J.J.'s temperament was standoffish. J.J. paid little mind to the millions of visitors on the other side of her glass-sided 1.7-million-gallon pool—even if those visitors were awestruck and deeply engaged at the sight of J.J.

But J.J. was physically affectionate with her caretakers. "It's as if she understands what we mean to her. She's bonded to us," the park's chief veterinarian, Tom Reidarson, said. Five months into J.J.'s rehabilitation at SeaWorld, the staff made the difficult decision to detach from her to make sure she was not so deeply imprinted on humans that she could not survive on her own in the wild.

Because so little is known about gray whale acoustics, senior researcher Ann Bowles at Hubbs/SeaWorld Research Institute studied J.J.'s vocalizations. When she played whale sounds inside J.J.'s tank, Bowles could study the captive gray whale's response, as well as introduce J.J. to whale vocalizations she would encounter in the open ocean. Bowles discovered that J.J. already moaned, croaked, and grunted like adult gray whales. What she had to learn was the audible "popping" that scientists theorize these grays use for navigation. J.J. was also studied during sleep by a team of Russian scientists, who found the first evidence that these whales dream.

In her brief captivity, J.J. was at least safe from predators, but would she learn to fear and defend herself from orcas in the open ocean? To test J.J.'s response to captive orcas, researchers opened up J.J.'s tank to the nearby orcas, such as Shamu, via an underwater gate. Immediately J.J.'s behavior showed agitation. This protective instinct would serve her well in the wild, especially when she journeyed past deep, submarine canyons in Monterey Bay where transient orcas often migrate, preying upon gray whales.

Many researchers learned new aspects of gray whale biology and physiology from studying J.J. during her terrestrial life. When we first visited the Monterey Bay Aquarium in April 1998, the spring gray whale migration was in full swing. And just as there were many friends along the way for the gray whales, there were also many enemies. Newspapers that month chronicled a life-and-death saga of two calves killed by a pod of orcas in a six-hour ordeal. The orcas encircled their young prey, ramming and internally injuring the calves. "The strength of the calves is incredible," marveled one orca researcher, Nancy Black of the Oceanic Society. "It's pretty brutal—you feel for the calf."

The gray whale mothers are powerfully protective, sometimes even carrying their calves on their backs to hold them out of the water and away from the mighty orca teeth. Black has observed these mortal orca-gray whale encounters for years and noted that when the grays are closer to shore and not vulnerable out in the deep waters over the Monterey Canyon, they may find refuge in shallow waters or rocks where orcas will not follow. No one knows exactly how many gray whale calves die each year as prey to orcas. Yet the grays have their own considerable defenses, evolved over many migrations and millions of years of adaptation.

This ability to adapt to a changing environment is one of the survival skills that the gray whales model for us. As the whales rebound along the West Coast, scientists look to them as an "indicator species" that shows us the health of our shared ecosystem. As the gray whales are restored in greater numbers to our ecosystem, what does their return restore in us? Could it be that by more generously sharing cultures and habitats with the gray whales, their health can teach our species something about balance?

Vicki Floyd, a marine biologist working as a naturalist for Baja Expeditions in San Ignacio Lagoon, studied the lagoon's winter population of grays to discover patterns of vocalization, social communication, and mother-calf bonding—all essential survival tools for any species. She was also one of the researchers who worked to help rehabilitate J.J. and ready her for release into the wild.

"J.J. gave us keys into what was instinctive in gray whale newborns and what was learned. From the time the calf was three months old, we'd play for her other gray whale vocalizations recorded in the birthing lagoons and the Chukchi and Bering Seas," Floyd said. "Then we'd reinforce these vocalizations with food so she would associate gray whale sounds with feeding. Upon release, we hope she will follow gray whales and find her own food."

Floyd told the story of finding J.J. in the spring of 1998 floating always on one side of her pool. Researchers wondered if J.J. was lonely or depressed. Were there not enough stimuli in this tank, which by now was almost too tiny to contain her massive nine-ton weight? At last they realized—J.J. was always facing north! It was as if everything in her body was already pointing her toward her springtime northern migration to Alaska seas.

Floyd explained, "Necropsies reveal that a gray whale's brain has tiny magnetic particles—a black oxide of iron. We have evidence that these magnetic particles rotate and orient the whales along the Earth's magnetic fields."

As the time grew near for J.J.'s March 31, 1998, release, everyone was hoping that it would happen without a hitch. Coast Guard helicopters explored the seas off San Diego's Point Loma for signs of traveling whales in the hope that J.J. would fall into good company along that aquatic superhighway. Heavy equipment lifted and deftly lowered the huge, healthy whale onto the 180-foot deck of the Coast Guard ship Conifer as thousands watched and millions more witnessed on live television.

Now 31 feet long and weighing 19,200 pounds, J.J. had every chance for a successful release, if all went well. Scientists had outfitted her with high-tech gear—radio transmitters and several trained seals who had underwater cameras attached to their backs so they might swim with and track J.J. The hope was that whenever J.J. surfaced, the transmitters could be uplinked to a satellite, thus pinpointing her journey. Anyone could log on to the SeaWorld Web site and also follow her progress and survival. Her journey could answer many questions for researchers, as well. For example, no scientist has ever successfully followed a single gray whale along a migration from beginning to end. Every radio transmitter attached to a gray whale has been scraped off and left at the bottom of the ocean. Very little is known about the migration journey of juvenile whales, or why some grays stop along the way to become summer residents.

Scientists also don't know for sure if the Eastern and Western Pacific stocks of gray whales are separate groups or genetically linked. Scientists believe they are discrete, but there isn't enough DNA data to support this theory. We do know from photo-ID studies that Western Pacific grays have been seen in the Bering Sea and where the Bering intermingles with the Chukchi.

Ironically, the worst threat to J.J. was not orcas, but the humans who might approach her in boats—a flotilla of fans. "I just hope people will leave her alone," said Jim Sumich, whale expert and professor at Grossmont College in El Cajon, California. "She knows nothing of the ocean, she's going to make mistakes and maybe go the wrong way. She needs a wide berth."

J.J. was not only given the space in the open sea she so deserved after a year of captivity, she also managed to shed her radio transmitters and tracking devices within a few days of her release into the waves. As the world watched her body being lifted up from the Conifer deck and gently lowered into the ocean, many commentators wondered: "Will she know which direction to swim?" After a few wide circles around the boat, J.J. dove decidedly down and reclaimed her ocean life.

"It was remarkable," said Tom Reidarson.

"I'll miss her," SeaWorld's assistant curator of mammals, Keith Yip, said. "I'm sorry to see her go. We spent a lot of long days and sleepless night with her."

Reidarson added, "Some people have suggested it'll be like

watching your child go off to school for the first time. It's not. Your children come home again."

For a day or two, researchers tracked J.J. through the radio transmitters bolted into her blubber like a saddle on her knobby back. Designed not to harm the young whale, the radio packs were vulnerable to falling off, especially when she might be shoveling the ocean bottom for food. Several days after her release into the wild, J.J. was spotted near San Diego. She had turned north and was last sighted, before radio contact was lost, near a group of migrating gray whales. Her exact migration path was to remain a scientific mystery, because all our human high-tech tracking devices were probably at the bottom of the sea. Nevertheless, J.J. was the largest animal ever successfully released back into the wild—and the precedent she set for the future of wildlife rehabilitation was inspiring. She will never be forgotten. Some of us still scan the waters every spring and fall, hoping to see J.J. rise up from the waves and carry on with her long life in the sea.

It is as if the gray whales are both familiar and alien, their world open and closed to our scrutiny. No one knows the fate of GiGi or J.J., whether they died in the open sea, were hunted along their way, or became some of the Friendlies along the Coast who continue to charm and engage millions of whale-watchers.

Perhaps J.J. has come home again, but no one has recognized her. Perhaps in Baja lagoons she is being watched and even touched. Much of the gray whales' mystery comes from their anonymity. Even these transient "superstar" gray whales, the "charismatic megafauna," as some scientists call these species that capture our human imagination, still disappear again into their own.

Following the Sun

The gray whales must migrate by following the sun, summer to winter; its radiance is another part of the map of their lives that powers their journey north or south. It becomes their destination. By necessity, it is their god, their provider and guide. As their provider, the sun fuels the storms of plankton blooming in its light, making rich feeding grounds for the whale. To follow this food source, the floating world, they journey toward the longest hours of daylight, as the plankton flower and increase because of the light. They follow plankton and plankton follow the sun.

Unlike other whale species, the gray whales are locked to the land. Not only must they surface to breathe, but they live, by necessity, in the sunlit habitat of coastal waters where plankton are rich. It is unfortunately also the place where humans travel, this rich mixing zone where land and water meet. This is one of the ironies *of nature, that they, so large, are also so vulnerable. They are both nurtured and damned by their need for light. This vulnerability is why the gray whales are now extinct in the Atlantic and nearly so in Asian waters. It is also why they have twice been brought to near extinction along the Pacific coast. Their food needs and their long migrations expose them to enemies along their way, primarily people, but in much smaller numbers, killer whales, and in the past, at least, swordfish. Some Native hunters and others had the vision to cease hunting and those who named them as an endangered species.*

Keeping Count

Mentally mapping the mighty gray whale migration requires a leap of the imagination. If this passage of thousands of mammals happened on land, instead of at sea, it might be hailed as the epic journey it truly is.

Sometimes the magic and majesty of other species eludes us until we actually imagine ourselves part of their mysterious lives. This is why education and stories about other animals' lives are as vital as conservation and scientific study. And volunteers along the California coast tell some of the best stories of all.

Cabrillo National Monument, high on the cliffs of Point Loma in San Diego, is a Mecca for southern Californians who, while going about their lives in the urban wilds, also want to experience the great, gray voyagers swimming right past huge cities of humans. San Diego may seem like the far southwestern tip of America, but to the gray whales it is already approaching the midway point in their northern journey.

At the monument is a lighthouse that offers whale-watchers telescopes and a glassed-in observatory. When we visited on a sunny day in late December during the southern migration. The building was full of delighted witnesses. Some were there just for a visit; others were keeping count.

These census takers count whales throughout the daylight hours and estimate nightly numbers of passing grays by multiplying by two the whales counted during the day. Keeping count requires such patience and humility that the public often overlooks such a seemingly simple, yet crucial, part of volunteer conservation and science.

For marine biologist and census taker Alisa Schulman-Janiger, counting gray whales is a passion. Schulman-Janiger is a marine biology teacher at the Marine Science Magnet of San Pedro High School, near the Los Angeles Harbor, and is one of the instructors for the Cabrillo whale-watch class. She has spent much of the past 20 years conducting whale photo-identification and behavior studies, specializing in gray whales, killer whales, and humpback whales, and has served during this period on the board of directors for the Los Angeles Chapter of the American Cetacean Society. She has been the director of the ACS/LA Gray Whale Census and Behavior Project for the past 17 years. Begun in 1978, the volunteer project has run annually since 1984. In addition, the National Marine Fisheries Service has conducted an official gray whale census count from the Monterey area nearly annually since 1968. During both the fall and spring migrations, approximately 45 core participants work in shifts of two to four persons, watching and counting whales from sunrise to sunset, seven days a week. Their census vigil begins December 1 for the fall, southern migration, with most whales sighted in late January and early February. The census count for the spring, northern migration begins in late February and extends until May 15. The count represents seven months of diligent, devoted work for the humans, and difficult, dangerous passages for the gray whales.

The Census Project is not just about counting gray whales during their massive migration and tracking the percentage of calves. It's also about observing gray whale behavior, seeing how close to shore they migrate, when the biggest and smallest "pulses"—or waves—of whales occur each season, and what other species of marine mammals use these waters. There is so much to study that the long days seem short, says Alisa Schulman-Janiger, in a voice that reminds me of my favorite biology teacher in school. She is calm and informative, but her excitement about these grays is obvious.

"My favorite day with the gray whales?" she says. "It occurred during the census in April 1994. We watched a very playful gray whale calf breach 64 times as its mom swam nearby just outside the surf zone, constantly breaching for as long as we could see it." Such high-spirited acrobatics are not the norm for gray whale calves. Schulman-Janiger explains, "The typical number of breaches is more like three or four. But a sea lion cruised by this calf and then they started swimming in circles together. The mother gray actually had to come back and physically corral the calf to continue their migration—her calf was having so much fun! We still talk about that day."

Volunteers have noted that they sight fewer whales during the southbound migration because grays swim farther away from shore here than in the spring, when they tend to hug the coastline. Perhaps this spring intimacy is because there are so many mothers with newborn calves going north, and the shoreline's coves and kelp beds can offer areas to nurse, rest, and be sheltered from potentially deadly weather and hungry killer whales. Kelp forests are one of the most important and productive of all ecosystems, offering food, safety, and habitat for hundreds of different fish, algae, and invertebrate species.

In the late 1980s, Schulman-Janiger notes, the Census Project's data on near-shore calf sightings helped contribute to the passage of a California state referendum to ban gill-net fishing within three miles of shore. "You don't know how a species is truly impacted by something unless you know what its migration timing, distribution, and behaviors were before," Schulman-Janiger explains. "So you need to collect baseline data." The Census Project's data is used to monitor long-term trends in the gray whale population.

Many obstacles threaten the gray whales on their long journeys, says Schulman-Janiger. "We have problems with high-speed vessels in California, where many boaters are not aware of whales or of guidelines for whale-watching that would keep them from illegally harassing the whales. It's difficult to enforce the 100-yard distance guideline or other anti-harassment suggestions. We have a big problem with Jet Skis and shipping traffic." Schulman-Janiger tells the poignant story of a gray whale mother hit by a big ship at 20 knots, right in front of hundreds of whale watchers. "There was so much blood in the water," she recalls. "The whale-watchers were helpless, just had to witness this horror until at last the mother just sank. Who knows what happened to her calf?"

Such collisions between the world of gray whales and our own human traffic are too frequent. Witnesses who can document these events are invaluable if we are ever to educate ourselves and set up guidelines for coexistence. Some researchers are even suggesting a "Whale Lane" along the migratory routes during peak seasons.

For Katy Penland, President of the American Cetacean Society, an encounter with a whale changed her life.

Penland was photographing pelicans in the fall of 1992 off a dock near the Palos Verdes peninsula in southern California. Focused intently on her camera angles and shots, she did not notice the curious creature just beneath her dock in the shallow waters. Suddenly she found herself gazing through the viewfinder at a huge shape cruising along the pilings.

"Just like a big gray basketball with lumps on it," Penland said, laughing. "That's my first impression of this whale right under my feet. I ran alongside it, following along the pier above this critter. It was about 20 to 25 feet long and I found out later it was probably a juvenile on its very first migration south. You know, those juveniles, they explore every nook and cranny along the shore. I think their mothers say, 'Keep the coast on your left all the way down. And swimming north to Alaska, keep the coastline on your right.'"

Penland marks that day eight years ago as the beginning of her passion for gray whales and all other cetaceans. As she ran delightedly after the juvenile gray as it headed out to Rocky Point at the south end of Santa Monica Bay, she could not have known that her photography would find a new meaning. As a volunteer naturalist, Penland now photographs whales for identification, adding immensely to the scientific documentation and understanding of other species.

"After that first encounter with the juvenile gray, I looked up 'whale watch' in the white pages of the Redondo Beach telephone directory. I was hooked! Every day I went out whale-watching that whole season. People wearing red jackets covered in patches were talking on the boats, sponsored by Whalewatch, a volunteer naturalist training program. They train around 200 people a year to do the narration on the commercial whale-watching boats and to teach about gray whales in local classrooms and to civic groups."

Katy Penland calls the hundreds of volunteers all along our West Coast who count and watch whales "eco-heroes." "A day in the life of a census taker like those who work with Alisa is rigorous," she adds "They get up at dawn and count until dusk. People must be trained well—and all this is volunteer work, sometimes in addition to a full-time job! For example, Alisa is a full-time schoolteacher. And there's another guy, a retired pilot, who uses those sharp eyes to keep track of this super marine highway. He's logged thousands of hours in this land-based study."

This volunteer, retired TWA pilot Jud Goodspeed, is aptly named. Every day during the census he takes his post in a lawn chair with binoculars. Beginning at high noon, he watches whales until 5 p.m.—rain or shine. Such commitment to the natural world is a fine way to live, as pioneer author Willa Cather had inscribed on her tombstone, "that is happiness; to be dissolved into something complete and great."

Penland says almost wistfully, "What we really know about these marine mammal stocks is so small. We've only been formally studying them for the last 30 years or so—which is only half the average life span of a whale. Prior to this, everything we knew about whale population figures came from whaling records."

Her solution? Keep counting. Keep track. Keep an eye on the whales who share our shallow waters and seas.

Forever Changed

In my own life, I went to the water and was forever changed by the eye of the whale in which I saw an entire world. Peace is here, in accepting that smallness of our place in the world, in being as humble as the birds around us. This humble place is where we belong, according to the tribal view. Like the whales in the ocean gazing from the kelp beds and seeing the turning stars at night, when young people look at the clouds, the movement of an ant, the wings of a moth, or the leap of a fish—they are part of it.

But few people think of what is already missing, that there is a hole not only in the world, but also in the universe we used to call infinite, once believing it was without limit. This is true with the gray whales in the Atlantic. Do we now miss them? The Earth is yet an unknown mystery and we have barely begun to understand this.

In centuries past, Chumash lands extended over a large area in the region near Santa Barbara and Santa Monica, and included the coast and its islands, as well as the Channel Islands. In early times, the Chumash had beautifully constructed canoes and an elaborate culture. Rock paintings, cave paintings, and buildings remain, also items as small as exquisitely made baskets, nets and twine. The Chumash had a world in which balance was significant. The entire world and universe, according to their calendar and cosmology, were, like water, in constant motion and flux. With their knowledge system they knew there was no separation between humans and nature. Each Native tribe still speaks of this participatory relationship. The human is only a part of all the rest.

Honoring the Dead

The notes of ethnologist Reverend Bowers were not always dependable, but some of his data about whaling are confirmed. At Chumash sites he recognized cemeteries by whale bones, including ribs, scapulas, and sometimes bone slabs. They were grave markers. He also found two mortars, used for grinding grains, that were carved from the vertebrae of whales, and he took one with him. This particular mortar measured ten inches in diameter and was "beautifully rounded at the bottom." He wrote, "It is a fine specimen and unique as far as I know." Other writers and anthropologists also mentioned rib bones over the bodies of the dead, as if the whale spirit honored the dead. Whale-bone markers were found not only on the coast or islands in the area, but also further inland, even in a village as far as 30 miles from the coast.

Bowers collected numerous items, and though he may have been a grave robber, he did gather some significant information. He *said that the last survivor of Santa Rosa Islanders told him that his people worshipped the sun because it gave them warmth and light, and the swordfish because it killed the whale, which washed to shore, and the killer whale for the same reason, because it furnished them with food. They waited. Whales arrived. The washed-up gray whales were welcomed.*

Sightings, Skeletons, and Blessings

Every morning during the spring and fall, people along the coast from Oregon to Victoria, British Columbia, open their e-mail to find these messages: "Listened to the grays sighing all night. Better than star-gazing!!" or "Gray whales traveling south along Saratoga Passage—headed your way..." or "Gray whales feeding on our island's ghost shrimp beds. Rolling in the muck. Maybe they'll stay with us awhile?"

Every spring and fall the Whidbey Island, Washington, Sightings Network is a-flurry with phone calls, e-mails, and natural history questions as people watch the gray whales migrate past their everyday shores. These daily alerts during the gray whale's migrations stream in from over 300 participants in the nonprofit Orca Conservancy Whale Sightings Network. The Internet subscribers' list to this free Sightings Network includes the National Marine Fisheries, media, researchers, marine educators, politicians, volunteers, and whale enthusiasts. Sometimes the responses to the Sightings Network come from as far away as Japan.

The Whale Sightings Network's stated purpose is to "involve citizens in helping researchers track the movement of whales and encourage people to observe whales from their homes, businesses, and beaches." These daily sightings are not only a link between whales and people, they also inform and educate. A typical sighting from Susan Berta, who coordinates the network: "April 19, 2002: We didn't see any grays today, though at low tide the beach below us showed the extent of the feasting that's been happening in this neighborhood lately—feeding pits virtually covered the entire sandy beach! In fact, the tide is probably up now, so I'd better get out and listen in case they're out there."

To include the gray whales in one's definition of "neighborhood" is to reach out and embrace another species as extended family. This is what the indigenous peoples have always done in their prayers, which include "all my relations." For those of us who receive these sightings, the sense of connection between people and whales is a daily reminder that we are not here alone, but part of a web of life, a kinship system between species.

"What unifies all these different people is a true love and passion for whales," says Susan Berta. "We've been astounded by the increase in sightings and gray whale information that helps researchers understand local whale behavior, as well as see the bigger picture gleaned from sightings over a large geographical area."

Berta and Howard Garrett, both longtime researchers and activists associated with the Center for Whale Research on Washington's San Juan Island, are at the hub of all this Sightings Network activity. They serve as the nerve center for the incoming sightings of diverse cetaceans. Sometimes the reports are from as far away as Vancouver Island; some are from as far south as California. Other Internet subscribers include East Coast dwellers; although the gray whale is extinct along their Atlantic shores, these people enjoy tuning in to the daily West Coast whale bulletins. "It's like watching the weather report for the entire country!" says a Sightings subscriber from Philadelphia.

Of particular importance are the dozen or so sightings each month of the resident orcas of Puget Sound. Named the J, K, and L pods, these three pods make up the Southern Resident Community of orcas. As resident orcas, they do not, like transient orcas, feed off gray whales, but instead subsist on a diet mostly of herring and salmon. The Sightings Network was also invaluable in helping increase awareness of the 364 gray whales who died in 1998 and 1999 along the West Coast migration route. Berta notes, "The die-off along the Pacific coast in 1999 was twice as high as any year dating back to 1985, and the highest in the 24 years records have been kept."

One of those starving whales washed up off Whidbey Island in December 1998. Some people might consider a 32-foot-long whale carcass an eyesore and a burden, but not this active community. Dead gray whales are considered the responsibility of property owners to dispose of—and in ignorance, some people have tried to rid their beaches of the thousands of pounds of decomposing blubber by using everything from dynamite to chain saws. But thanks to the education and community of Sightings Network subscribers on Whidbey Island and the very active Island County/WSU Beach Watcher volunteers, this gray whale was given a name, Rosie, after the scientific term for gray whales—*Eschrichtius robustus*.

Instead of burying or towing her back out to sea, 40 Whidbey volunteers, from children to senior citizens, waded into the shallows and worked for four cold days to remove hundreds of pounds of blubber and entrails before salvaging the skeleton.

"We called Rosie our Christmas whale," says Penny Bowen of the scores of Whidbey Island volunteers who literally put a gray whale back together again.

Penny stood inside Rosie's skeleton as it lay on the mud flats in January 1999, strewn with some semblance of a whale's outline. In awe and with a certain reverence, we walked slowly among Rosie's 28 rib bones, paused to ponder her gigantic beak-like skull, the large fan-like finger bones from her pectoral fins and the tiny primitive leg bones still embedded in the skeleton. These rudimentary bones are the nubs of hind legs that shrank as the whale evolved for life in the sea. But their fossil evidence reminds us that these ancient cetaceans once walked on land. I marveled at the fact that these mammals had gone back into the ocean after their terrestrial life, just as humans had evolved from the ocean onto land.

Then Penny told me the story of how Rosie brought a town together.

"It was so great to see the kids learning about gray whales," Penny says. "When we were stripping the blubber, some of the kids were jumping up and down on the huge whale's tongue like a trampoline. The coils of intestines were amazing. One of the ten-year-old students said, 'This creature is teaching us.'

"So here we all were up to our elbows in whale guts and the smell was almost overwhelming. Like seaweed and oil. It doesn't smell fishy or like spoiled meat. It's kind of an overpowering, spicy, pungent aroma. It clings to your skin. I have to say it's not a pleasant scent," Penny admits.

After flensing (stripping the blubber from the whale), the massive skeleton was taken to the Seaplane Base for more cleaning. Rosie's bones were stored in perforated 55-gallon plastic drums that were then sunk underwater and tied to a dock in Crescent Harbor. Rosie's 6-foot-long, 400-pound skull was strapped to a wooden pallet, covered with chicken wire and anchored underwater to let sea stars, crabs, and other small marine life diligently clean the bones. At last the cleansed bones were left to bleach in the sun for two weeks, then sanded carefully by hand. The plan was to rearticulate the gray whale for educational display on the Coupeville wharf.

"We did the job so that everyone else can share in seeing this wonderful creature who symbolizes all the grays," Penny Bowen concludes. "After all, the grays are our island neighbors."

Two years later, on Easter, 2001, I visited Rosie again. This time, her elegant and seemingly endless skeleton was poised midair, perfectly remembered, and hanging in the Coupeville Wharf Harbor.

"It's kind of a rebirth," Susan Berta says proudly. "I think Rosie would be honored to know that even in her death, she lives on to bring the wonder of whales to all who see her."

There was yet another wonder for the Whidbey Islanders to discover about their reassembled gray whale. DNA testing revealed that Rosie was male—so Rosie evolved into Roosevelt (or "Rosie Grayer," like Rosie Greer, the athlete). Gray whale expert John Calambokidis announced this discovery at an Orca Network meeting in April 2001. Calambokidis and his Cascadia Research Team have been studying and cataloguing gray whale photo identifications since 1990, and his scientific work has been invaluable for increasing our knowledge of gray whale behavior and biology.

Calambokidis has identified at least five individual gray whales who return year after year to Whidbey Island to feed on abundant ghost shrimp beds. "This is a stable group of seasonal residents," explains Calambokidis. "They are part of a larger resident group of grays, about 200 to 250, who linger along the migration route on the Washington coast for a three-to-four-month period. They're usually gone by May or June. We don't know where they go from here. We do know that these seasonal residents practice what scientists call 'site fidelity,' returning to the same territory year after year." Calambokidis theorizes that mother whales teach their young site fidelity on their first migration, and after that the offspring remember and return as residents themselves.

At a lively and packed meeting in the spring of 2001 at the Whidbey Island church, Calambokidis answered questions from islanders eager to understand and protect the five resident gray whales circling Whidbey Island, feeding on seasonal ghost shrimp.

Most of the questions had to do with gray whale deaths and the many dangers along their migration path. "Our records go back to 1977," Calambokidis notes. "There has been an average of 4 strandings off Washington State per spring migration, with a peak of 14. But what happened in 1999 and 2000 couldn't be put on the same graph as any previous year. In 1999, 26 gray whales died on shore, and 23 in 2000. That was in Washington State alone. And it also happened all along the West Coast from Baja to Canada. Most were animals who had starved. Necropsies revealed that the percentage of whale oil left in the blubber was down to only one to two percent. The number of calves born plummeted to the lowest level ever seen—only a few hundred."

He discussed the possible causes of this unprecedented die-off. The main cause is most likely the decline in ocean-bottom amphipods, which grays feed upon in the Bering and Chukchi Seas. This could be due to the combined effects of rising sea temperatures, decreased ocean productivity, perhaps a viral infection, ocean pollution, or the possibility that the ocean has reached its carrying capacity for gray whales. "Carrying capacity" is the term scientists use to describe the maximum population level at which a species can be sustained, or carried, vis à vis its finite food supply. In other words, can the ocean bottoms continue to supply enough nutrients to ensure the gray whale's continued survival? One thing is certain: The grays are starving to death. Their emaciated carcasses are an alarming sign.

Another concern for Calambokidis, as well as for many Washington State residents, is the renewed Makah whale hunt and the precedent it sets for other tribes who have signaled their intent to return to whaling, as well as its impact on the gray whale's future. "Originally the National Marine Fisheries Service evaluated the Makah hunt strictly on a rebounding population of 26,000 gray whales," Calambokidis explains. "They didn't take into account the difference between migrating and resident grays. If the Makah's seasonal take is five animals out of a resident population of 200 to 250, and if the other Canadian tribes eventually return to whaling, this could mean 15 to 20 whales per year out of a much smaller resident population of 250, instead of the migrating 26,000."

"The good news this year," Calambokidis adds, "is that as of May 2001, we have not yet seen the die-off of grays that we witnessed in 1998 to 2000. But it's early yet and the mortality of grays is usually highest in late May and June when the stragglers try to finish up their long springtime northern migration."

Still, no one knows the reason for the terrible die-off of grays whales in the springs of 1999 and 2000 or why there is not the same significant death rate this spring of 2001. The very latest census of gray whales shows a decline in population to 17,414 in May 2002. This dramatic drop from the 1997-1998 high of 26,000 alarms many environmentalists. What will the next years be like for the gray whales? Will there be more hunting, more die-offs, or will they thrive alongside us?

A World of Other Species

Humans are born as the very best, with beauty held within. All around us is a world of other species, magnificent with grace or light or strangeness. When there was a gentle wind that moved the sailors of all the worlds, perhaps it reminded them of the same loving sounds as the whispers of their mothers or wives. Yet a man can still open like a seed and close like a lock. The mind can do things the heart can't follow. The body can exact deeds the soul rebels against. The lives of men from a European shore far away can trouble an entire world; yet anchored here, crossing the boundaries of human skin and whale, there is still some tenderness inside each person. Perhaps we have long had the stirrings to become something other than a human. But being part of that means we may kill what we also love.

Brother Leviathan, Sister Whale

For some, conservation is also a matter of spiritual practice. As the gray whales make their arduous journey by Whidbey Island and other communities on their way to Alaska, some people celebrate their passing not simply with whale-watching and marine education, but by actually saying daily prayers for them.

While this spiritual kinship is not as integrated into our daily society as education and science, it is a natural outgrowth of the burgeoning eco-spirituality movement: an interdisciplinary ethos that bridges science, spirituality, and environmentalism.

In a recent issue of *Earth Letter*, published by the Christian environmental stewardship group Earth Ministry in Seattle, Washington, Episcopal priest Jim Friedrich tells the story of an interspecies worship service with gray whales and humans. Friedrich, who produced the popular video, *The Greening of Faith*, lives on Whidbey Island, Washington. While visiting in southern California, he was asked by the Natural History Museum in Santa Barbara to perform a "ritual improvisation...the blessing of the gray whales who pass by every March on their way to Alaska."

Originally the Reverend Martha Siegal, a local priest, was to perform the ceremony, but she couldn't find the proper blessings in her theological texts. "There are Anglican blessings for crops, homes, pets, and automobiles," writes Friedrich, "but not for whales." So Friedrich composed his own blessing. As priests, congregation, naturalists, politicians, and the news media boarded a boat for this "interspecies ritual," no whales were in sight. But as the blessing began, a group of gray whales suddenly swam alongside, delighting the celebrants. Friedrich read his blessing out over the waves and the grays:

Brother leviathan, sister whale, dwellers of the deep,
We greet you in the name of the human people
Who share this planet with you.
We stand in awe of your powerful beauty,
We take delight in your liquid dance,
We rejoice that another spring
has brought you to grace our coastal waters.

May the Creator, whose breath causes all things to be
Keep you on your journey home,
Sustain you through the summers and winters to come,
And make your very existence a song of praise and wonder....

"When the words of blessing were uttered," Friedrich writes, "the whales began to breach and spout, as if praying the responses appropriate to their kind. Human cries of delight completed the antiphony, and the whales departed in peace."

Jane Goodall has written, "That which is loved, I believe, can grow." There is much growing to do in this new century; there is much to hope for between our species as the gray whales find more friends along their watery way.

This contemporary blessing of the whales echoes other indigenous songs and prayers for the whales and their traditional passing along our coasts. The ancient bond between whales and peoples is only growing stronger as we enter a new century. One wonders, Will our grandchildren know more about the mysteries of these gray whales? Will the reverence of the humans and the friendliness of the whales flourish for many more centuries?

PART THREE

Double Vision: Hunting the Whales

Makah Reservation, Neah Bay, Washington

1996 to 2002

Greeted With Reverence

"When I was young the whales came up and they used to scrape the reef to get their barnacles off."

ALBERTA THOMPSON, MAKAH ELDER

In the Northwest, men once asked the whales to give themselves. It was all very different then. The whales were greeted with feathers and songs.

Soon there were no more whales, and men did not ask a whale to offer itself, and the world, it seems, was changed forever. But forever is too long a time and so I say only "it seems," meaning there is hope, and I do hope.

I think of our debt to the animals. We depend on them. As Native peoples, especially. The one thing most indigenous cultures had in common was a respect for the world upon which they depended, knowing their own place in the surrounding world. Ceremonies of gratitude were always a part of the traditional ways. A whale would be addressed, hunted, and greeted with reverence.

The whales now show us their young, gaze at us with ancient eyes, rising from the water, sometimes carrying their infants on their *backs, lifting them up before us in their greatness and homeliness and even their holiness. For once in a deep span of time I was one with you, they say, perhaps one of you. Look at my hand and legs and hipbones when you see through or into me. Once when we died at your hands, none of it was for money and should never have been, for we were great in the oceans for something more fine than that, more right than that. Here is my infant. Look at how I present her to you before she is even covered with barnacles, as she is fresh and dark and shining when I first bring her upward and present her to you.*

As the water falls from them, as the eye opens and sees us, as the knuckled back rises with the baby alongside it, I hate for a moment being human and what we do and I wish I could cross the boundary, cross the world to an older self. Underwater and in kelp beds and primal silt is a language and life never to be deciphered by mere men.

All this is part of the new relationship of humans and whales, and I know what the gray whales face in their journey north and I try to put it out of my mind. This year they will pass the hunters.

The whales' friendliness in the lagoons of Baja has helped them, has changed people, making them fight against hunting rights and salt factories. It has also made the whales vulnerable. I think of my responsibility as a Native woman to preserve the world for the future. It is also my responsibility to stand up for treaty rights. And here is where the conflict begins.

Treaties do not hold in water and nature; there are other laws at work than those on paper. Not just spiritual laws, not just treaty rights, the world around us has its own law. For now there is the law of ocean, water, lagoon, and whale.

As Iroquois elder Oren Lyons said, "Who speaks for the eagle, who speaks for the whale?"

There is already a history with Indians and whales. In the 19th century, Native peoples helped Europeans and other nations whale, then mourned the animals' extinction and the accompanying cultural losses. It was a time of great changes. The Europeans also wanted fur. The killing of sea otters and other furred animals added to the losses. Then the smoke of Russian, British, Chinese, and American whaling ships disappeared, leaving even more losses behind them. The animals were gone, the whales were nearly extinct, and the whalers from other places had vanished. The great abundances of life diminished to a trickle.

In 1995 Brenda and I drove through the deforested world near Neah Bay, on the Olympic Peninsula in the northwestern tip of Washington State. I said, "It must have been beautiful," because it still is, even logged. Driving, I thought of us as tribal peoples. We, too, were—sometimes still are—beautiful with our ceremonies, our clothing, our ways, and our compassion, even empathy, for the food animals we hunted.

We were traveling to talk to a few older Makah women about their protest against the tribe's planned return to whaling. I felt a great question and conflict inside myself. I support treaty rights. We are sovereign nations as tribal peoples, and I have always been first to stand up for those rights. But I also support the life of a great creature who is still endangered, with compromised health and in a compromised ocean. I oppose killing it if it is not necessary, not done with ceremony in the aboriginal relationship with the whale. I support the hunt in waters farther north, where tribes have continued in an unbroken tradition as whalers. But there are other indigenous whalers, Siberian, who used the whale meat for fox farms, in order to sell the fur for coats. This is not a part of the sacred agreement, as in the hunts where the whale is welcomed, sung to, and given back to the sea with prayers.

As Natives, we need to hold to our wisdom and traditions, and they include an agreement and relationship with the animals; their souls meet ours, spirits who know each other in ways interminable.

A New Hunt

As a new century begins, an old whale hunt returns. The Makah tribe, which stopped whaling 70 years ago, is planning to return to that traditional practice. Feelings are running high among the many parties involved, including animal conservationists and activists as well as pro- and antiwhaling forces within the tribe itself. Amid the media glare, tense polarities, anger, and blame on both sides, there is another story that is not being told about the Makah tribe's return to whaling. We are setting this story down here as living history with hope for a vision that sees beyond today.

Three and a half hours from Seattle, on the windswept far northwestern tip of our continent, the Makah people have made their villages for centuries. Thousand-year-old Makah petroglyphs of whales face-to-face with round, wide-eyed humans show that the tribe's survival was interwoven with the gray whale's migration. Native hunting was a way of life for the Makah, and its bounty included grays, sea otters, humpbacks, seals, and sea lions. It was a rich harvest in a world of seemingly limitless natural abundance. For over a thousand years, the Makah were one of the most flourishing and powerful of all Northwest tribes. Their potlatches—rituals of gift-giving—were among the most generous. For generations, through ceremony, art, and hunting, the Makah honored the passing grays. Whale hunts were held not only for subsistence, but also to seek spiritual balance with other species and the natural world.

Ironically, the history of the Makah and that of the grays bears some sad similarities. In the mid-1800s, at the same time the Makah found their villages ravaged by smallpox, measles, and other diseases brought by European settlers, the gray whales were being slaughtered by Yankee whalers. Both Native peoples and gray whales faced such terrible losses that some observers wondered if either would survive the rapacity of European conquest. After Yankee whaling brought the grays to fewer than 1,200 individuals at the turn of the 20th century, the Makah tribe was the first to stop whale hunting. The Makah elders' 1915 decision was a farsighted example that was not followed by other nations until the International Whaling Commission (IWC) in 1946 prohibited the hunting of gray whales.

Though the Makah elders voluntarily stopped whale hunting early in the 20th century, they never gave up their treaty right to hunt whales, as did some other tribes. In the early 1990s, along with 14 commercial fishing groups and 19 other West Coast tribes, the Makah petitioned the U.S. government to remove the Western Pacific gray whale from the endangered species list, and the whales were indeed delisted in 1994.

In 1995, the Makah tribal council first approached the IWC with an official request to hunt gray whales. The reasons they gave included a need to return to their spiritual and historical traditions as well as a wish "to fulfill the legacy of our forefathers and restore a part of our culture which was taken from us." To meet IWC requirements that the hunt be humane, the hunters planned to follow up their harpoon strike with a killing shot from a massive .50-caliber antitank rifle.

The IWC also received a petition signed by seven Makah elders protesting the hunt and claiming the hunt would be conducted for commercial, not spiritual reasons. These tribal members believed that the hunters wanted to open the door to a potentially lucrative trade in whale parts with countries such as Japan or Norway. The elders' petition stated

> *[We] think the word "subsistence" is the wrong thing to say when our people haven't used or had whale meat-blubber since the early 1900s.... We believe the hunt is only for money.*

The Makah elders' eloquence at the IWC against their tribe's return to whaling deterred the Tribal Council, and they withdrew their whaling request that year. But in 1997 the Council again approached the IWC with a request to hunt gray whales. Amidst international outcry, the U.S. agreed to a trade of five bowhead whales from their Alaskan Eskimo quota for four gray whales from the Russian government, which receives a quota of 120 gray whales per year for Russian Chukchi subsistence-hunter Eskimos. The trade was not sanctioned by the IWC, and to this day, other IWC countries including Austria, Australia, and New Zealand, insist that, because this trade was not voted upon, the Makah hunt is illegal.

With the National Marine Fisheries Service, the Makah worked out a management plan allowing the tribe to "strike" a whale in the hunting pursuit 33 times over the next five years. The plan allowed the Makah to kill 20 whales, up to 5 per year, between 1998 and 2002.

The Makah Tribal Council began preparations for a fall 1998 hunt. Amidst international controversy and deepening tensions within and outside the tribe, a fragile coalition of moderate Makah and non-Native conservationists continued working together behind the scenes to find some common ground. Since 1995 this small coalition had worked to build a bridge between cultures and peoples—a bridge that was both pro-Makah and pro-whale.

When Linda and I first traveled together in 1996 to Neah Bay to answer the call of those elders who had been protesting their tribe's resumption of the gray whale hunt, we were responding to a direct request from tribal elder Alberta Thompson, who wanted to tell her story to writers who might see this conflict as more than just a polarized war. Thompson and several other Makah elders were worried that this hunt would divide her tribe and eventually weaken its spirit and her grandchildren's future.

So it was in June 1997, before the fall IWC meeting, that several moderate, undecided Makah leaders got together for a very quiet whale-watch in Neah Bay. Aboard the 54-foot boat *Discovery* were Tribal Council member Jerry Lucas; Alberta Thompson; her daughter Tracy; her grandson's wife, Amelia Davisson; and several representatives from nonprofit environmental groups.

With no other media aboard, and much good will among all sides, there was a sense of summer vacation. We all huddled against the early morning rain, what Jerry Lucas called "our liquid sunshine." For some of the Makah children, this was their first whale-watch. Our naturalist onboard was John Calambokidis of Cascadia Research. He, like Lucas, was keeping a discretely impartial distance from the media fray. Calambokidis is a gray whale expert, specializing in "resident whales"—those returning grays who seasonally linger and sometimes take up residence in shallow waters along the migration route from Neah Bay to Vancouver Island. Other whale-watchers included Dr. Toni Frohoff, marine mammal consultant to the Humane Society of the United States (HSUS), Will Anderson of the Progressive Animal Welfare Society (PAWS), and Stan Butler of the international group Whales Alive.

Dressed in bright red and yellow rain slickers, our group of 20 was a waterlogged but happy crew, especially Alberta's grandsons, who shared one set of binoculars. Together they scanned the horizon for signs of the gray whales' heart-shaped blow.

Jerry Lucas stretched his arms wide and called out over the outboard engine, "I want you to see what Makah take for granted everyday! Can you believe all this beauty?"

As far as the eye could see stretched luminous waves with roiling whitecaps. Eagles, gulls, and cormorants soared above our boat. Sea lions sleeping on buoys, harbor seals, and diving osprey brought life to the stony beaches. Close to shore rose the famous monoliths of Seal and Sail Rocks. This stretch of pristine waters and beach is the stuff of national parks, and the Makah are rightfully proud of their coastal homeland.

"We often see gray whales from shore," Alberta said, her face open, her smile expectant, "but this is the very first time in my whole life that I'll see whales from out on my native waters."

Within a half an hour from shore, Calambokidis spotted a gray whale. He directed our skipper toward flat-topped Seal Rock, encircled by a halo of seagulls.

"We've come full circle," Amelia Davisson said, smiling. She had accompanied Alberta to see the gray whales' birthing lagoons in Baja. "These whales we see here today may be the very whales we first touched down in Mexico."

Amelia and several other members of the Makah tribe were researching how to set up a Native-run whale-watching operation out of the new marina in Neah Bay. All over the world, from New Zealand to British Columbia, whale-watching has transformed depressed economies into flourishing whale-watching centers. With an unemployment rate of 50 percent, some Makah were thinking past the current whaling controversy toward ensuring a more secure economic future for their children. Other indigenous peoples had received government and corporate grants to sponsor what is now a multi-million-dollar business of whale-watching worldwide.

Neah Bay whale-watching could fill a void and meet the needs of many tourists who equate seeing whales with seeing the Northwest. "If the Makah ever decide to offer whale-watching, so many people would come here, their business would be an instant success," predicted Stan Butler. He pointed out, over peanut butter sandwiches in the boat's small galley, that whale-watching in Hawaii brought in a direct yearly profit of 30 to 35 million dollars, with an additional 150 million dollars coming in as indirect revenue from whale-watching spin-offs, such as hotels, car rentals, restaurants, art, photography, T-shirts, and sculptures.

But there was something unique to the Makah that Butler hoped the tribe would take into consideration in their attempts to revive their cultural heritage. "No one knows this land and these waters like the Makah," Butler said. "You have a cultural history with the gray whale that would give your own whale-watching more dimension than just a business venture."

"Can you imagine ecotours here?" Amelia asked as she talked about the possibilities of combined bird-watching, whale-watching, and a boat visit to the Makah's sister village of Nitnat across the Strait on British Columbia's Vancouver

Island. "We have so much of our culture to share. The Makah Canoe Club and Dance Troupe have gone as far away as Germany to share our traditions. There's camping here and fishing and so much history. Our carvers and artists are among the best in the world—if we can just get people to come up here to visit us and the whales."

As our boat rose up to meet gusts of wind and wave, one of the boys shouted, "Look, she's coming!"

Our boat tipped to one side with the weight of everyone rushing to see a mighty stroke of tail flukes slapping the water like a greeting. The whale breached before us and then dove under our boat. Suddenly there were three other whales surrounding us.

One sleek, gray-white belly streaked underwater right toward us. If I hadn't known it was a gray whale, I would have believed her a torpedo—fast and silver and massive. We all held our breaths. Five feet from the boat, the gray surfaced with a glide so her barnacled back was near our handrails. The whale rose up, turning slightly to make eye contact. A huge, dark eye opened as if to get a good look at these gap-mouthed humans. We exhaled in an explosion of shouts and high-pitched screams.

"She looked at me!" one of Alberta's grandsons shouted.

In our colorful slickers, we all leaned way out of the boat, calling to the whale. We had a sense of well-being, of all being well between us and the whales—a treaty and truce still unbroken. As if to underscore this contentment, the sun suddenly gleamed through gray clouds and within minutes we were peeling off our rain gear, surrendering to light and warmth.

"You see," Calambokidis said, "we didn't have to go very far to see gray whales. That's because these waters off Neah Bay are so calm and protected."

"You know," Jerry Lucas added, as he pointed out the jagged splendor of Slant and Mushroom Rocks, "I'd forgotten just how beautiful my own home waters are."

Though our coalition planned more Makah whale-watches, none ever happened. With the Makah Whaling Commission claiming success at that fall's 1997 IWC, the climate grew too polarized for any public meetings of those seeking common ground. Soon Makah whaling crewmembers were posing for the press with their antitank guns and harpoons; protesters of the hunt blocked roadways to the reservation. Yet in the late summer of 1998, against this backdrop of imminent conflict and potential violence, one of the whaling crew agreed to sit down and talk with two representatives of whale conservation groups. It was not a summit meeting; it was a conversation over a small, round table in a Makah home in what we would later refer to as our "kitchen table talks."

Micah McCarty is the contemplative young man many Makah expected to be "chosen by the whale" to throw the first harpoon in 70 years. He is also the great-grandson of the last Makah whaler, Hishka, whom his father, John, remembers as telling stories of the hunt. John McCarty's grandmother was also from a great whale-hunting family in Canada. "To be a whale hunter," John McCarty says, "you had to be born into that elite family lineage."

In 1855, when Washington State Territorial Governor Isaac Stevens came to the Makah villages with treaties, the tribe had 26 whaling chiefs. John McCarty, ex-director of the tribe's whaling commission, said, "Our chiefs were not interested in land allotments of 80 acres apiece like other tribes. We Makahs were interested in the sea, because we are People of The Ocean." The elder McCarty noted that if the tribe had kept its lands, it would have had trees and shore from Neah Bay all the way to Port Angeles. "Others have taken this land now and logged it for three generations," John McCarty said. "But we chose the sea and all its fish."

John and Micah McCarty spoke reverently about the spiritual aspects of the ancient bond between the Makah and the gray whale.

"We Makah are very concerned about the survival for the next generations of the gray whale," John McCarty said, expressing what many tribal members felt, but did not openly say to outsiders. "We want to do this hunt right, with dignity and tradition. Not to just kill a whale and have it float up spoiled on the beach. We have to do this hunt like the ancestors and bring that whale back up onto the beach for all the people."

Whaling Along the West Coast

From the oral traditions of hunters and storytellers, the story of the first whale hunter was given to well-known anthropologist Ruth Kirk by a whaler at Clayoquot Sound.

There were two chiefs. One gave lavish seal feasts superior to the other. Umik, the downcast man, dreamed that a kind man came to him and told him how to spear something larger than seals. He was told to purify himself in water for four days. And so he followed the directions, bathing. He was told to put his head underwater and rub himself with hemlock, which he did until he bled. "Imitate a whale, blow water," the man told him. He continued. He stayed underwater until blood came out of his ears and eyes. He heard a whale, as the man had told him. Following directions, he didn't turn to look at the whale, which lived where he bathed, in a lake.

He continued his teachings, remaining underwater until he vomited blood. Then, one day the teacher said, "Umik, I am the wolf, who has one heart with the whale of the lake." Umik was then permitted to look at the whale of the lake. It was small. He was told to keep it and take care of it. Then he was ready to kill a whale of the sea. This was the first whaler.

As they hunted during the fall and spring migrations, the people sang songs, towing songs, "Whale, turn toward the fine beach on the fine sandy beach." For indigenous peoples, not only were whales considered a food source; they were also beings of great power, creatures with a presence revered. They gazed at the sheer animal being of the whales with admiration and love, these creatures who sustained and offered life to them. Native peoples thought of the whales as part of their own nature. Their vision of the world was one in which there was a great coherence to all life, including an emotional and spiritual component. Whales, when landed by Natives, were greeted with awe and care. Songs welcomed the whale. Its body was covered with feathers, after having been hunted with so much respect and ceremony.

Based on the story of the first whaler, in a traditional hunt the whaling chief spent days singing to the whale and praying. He rubbed himself in order to bleed, then bathed in a spring, as if imitating the pain of the whale. His wife meanwhile sang, This is the way the whale will act, and she came toward the land, inland, moving toward the people. After the whaler left, the wife lay down, was covered with a cedar mat and sometimes with earth or enclosed in darkness, and she sent her thoughts that way.

As the chief went toward the water, he wore hemlock twigs tied to his forehead and a bearskin robe. He wore his hair tied in a knot and a ceremonial woven hat made by his wife, perhaps with a whale woven into it, a spear flying toward the whale. His was the head canoe, though there were others that would accompany this canoe, while more prayers took place as they traveled to the sea. Whale turn around and turn toward the beach... There was danger with the gray whale; it could jump entirely out of the water. It was a dangerous whale. It recognized its enemies. The mere humans knew what they risked for their food and oil and survival.

Whale I want you to come near me, so that I will get hold of your heart and deceive it, so that I will have strong legs and not be trembling and excited when you come and I spear you. Whale, if I spear you, I want my spear to strike your heart. Harpoon, when I use you, I want you to go to the heart of the whale. Whale, when I spear you, I want you to take hold of my spear with your hands. Whale, do not break my canoe for I am going to do good to you. I am going to put eagle-down and cedar bark on your back.

—*The whaler's prayer*

A Fragile Bridge Between Peoples

Neah Bay, Summer 1998

In the summer of 1998, a Makah whaler, young Micah McCarty, agreed to sit down and talk with two representatives of whale conservation groups in the hope of finding some common ground. That August, the whaling crew was seriously practicing every day in the canoe. Whale advocates stepped up their protests and the international media was encamped at Neah Bay. Amidst this volatile situation, our fragile coalition of moderates met, simply to listen to each other in John McCarty's kitchen.

John McCarty, ex-director of the tribe's whaling commission, and his son Micah, met with me, activist Ben White, and marine mammal biologist Toni Frohoff. We all hoped to share stories about the gray whales.

"What I learned about nature I learned from you people," Ben said. "You taught me to look in the animal's eye—there's a person in there. After I saw that, I couldn't kill animals. I had to protect them."

Micah was silent a long time. "So many of the old whaling songs and dances have been lost," he said. "We have tried to remember them and I want us to be one with the whale in spirit. I will tell you, that I don't know what I will do if, when we get out there and I look into the eye of the whale, I see that our crew is not in the spirit, not cleansed and ready in the old ways to take the whale—I don't know what I will do."

Then Ben told the story of living for three days and nights up in an ancient cedar in Washington's Dosewallips rain forest to protest the cutting of old-growth trees. "After you're up there in the treetops for days, you notice things...like how at dusk every evening there is this surprising shiver that runs through all the trees. You don't just sense it; you can see the trees tremble like with wind. Then someone told me it is the trees themselves going through their daily change—from breathing out to breathing in. And this kind of consciousness, that aliveness, we all share with each other and the animals as well."

Micah nodded, "Yes," he said, "we see the whales as *beings*. We even have ways of addressing the whales, and their ancestors, too." He continued, "At the time of the treaties [in 1855] it was only three years after a major smallpox epidemic.

A village of our people was obliterated. Stecowilth, his name means "gray whale," was concerned the government was taking away our way of life. They wanted to move us to the Cape and take up another way of life than our seagoing ancestors. Farming, industry. Stecowilth said, 'I want the sea. That is my country.' And so we got our fishing, seal, and whale-hunting treaty rights."

The Older World

The past is a country we look to as tribal people. Those of us who are indigenous want to find a way back to that past, to lay claim to the older world that sustained our ancestors with its richness. Our hearts still hurt for the injustice and corruption that took place at, and after, the time of treaty making, at the smallness and greed of the Americans whose life ways would destroy so much of our world and peoples.

In more recent times, we have become the ones others look toward. They seek us out as wisdom-keepers, as the first ecologists, as living examples of how to exist in the world. Yet many of us are also searching for a way back to traditions that were in place before the times of change. Some of us will find the way home, to the heart and soul, to indigenous knowledge. Some of us still have it. Some are too educated in Western systems of knowledge and values to return to the richness and strength that was the source, the brilliant *intelligence of our traditional cultures in the past and in many places and peoples, in the present.*

As an indigenous woman, I saw the story of the Makah and their request to whale as a tale bearing all the dimensions of an American tragedy. It became one. It was a story with many sides, one of which was eventually seen when the whale was killed.

Following the Ancestors

Makah elder Alberta Thompson was 73 in 1995 when she began a struggle of conscience with her tribe. She was speaking not only for the gray whales, but also for other elders. Dotti Chamblin, too, was outspoken against her tribe's return to whale hunting. Her great-great grandfather Ba-Ba-Sit, who died in 1907, was the last Makah to hunt whales. Raised in the Old Way of the Makah, Dotti is a traditional healer and professional in education and health care who ran for tribal council in 1996. Together, Thompson and Chamblin journeyed to the 1996 International Whaling Commission meeting in Scotland to deliver a petition, signed by seven Makah elders, against their tribe's proposed whale hunt.

"We were grandmothers arriving at the IWC in wheelchairs," Alberta Thompson said, smiling as she recalled the internal politics at the IWC. "The Tribal Council told everyone we were 'dangerous.' The Council went ahead and asked the IWC to go whaling, without the consent of the whole tribe," Alberta said. "They say they have 70 percent, but that isn't so."

"The fear of banishment from the tribal rolls by the Council has really stopped a lot of people from speaking out," Dotti Chamblin added.

More than anything else, these Makah grandmothers worried what effect this whale hunt would have on the future of the Makah.

As we listened to the elders bravely speaking out to both their tribe and the world, I realized that this dialogue went much deeper than treaty rights. It was about the connections we make among ourselves, our living world, and other species. These farsighted Makah elders—these grandmothers looking out for their children, as well as the great whales—speak for their own ancestors, as well as a species more ancient than our own.

Like this Native American tradition of balancing the subsistence both of body and spirit, any discussion today between anti-whaling environmental groups and the nations that want to exercise their historic treaty rights and begin whaling again must also be a spiritual dialogue. We must engage our ethics as well as our science and consider our future generations as well as our history. And in this debate, we must also respect the abiding culture and future health of another species: the great gray whale.

"The gray whales are the old ones of their kind," Makah elder Alberta Thompson told us that gray December day as we settled in her modest trailer to listen. "Just like us grandmothers."

~

Alberta Thompson and the other Makah women elders wanted to see their young people and tribal elders talk openly together about the relationship between the Makah and the great gray whales. Why, these elders asked, was there no open forum within the tribe for a whaling debate, and why was the Makah Whaling Commission not focused first on the spiritual training before any return to whale hunting, as always in the past?

Then Alberta shared with us a terrible vision. "There will be deaths," Alberta said softly. "Not just whales. My fear is that this whaling will divide and fragment our tribe. And that will backlash against our children. What is our future with the whale?" she wondered. "What if the tribe really does hunt again?"

It did. This pitiful tragedy contained the history of people who, by their early location as an easy trade center, and, as with many of us, by forced assimilation, lost their traditions. The whalers, before their next whale hunt, would tell the media that this first whale hunt enabled them to become sober. If the whale held that much healing, time will tell. But a year later, as we were leaving the reservation we saw written on a stop sign: STOP US. At first I thought it meant stop the United States, but it has other meanings as well.

When the White Men Arrived

When the white men arrived." This is the beginning of many Native stories that tell of change, from forces without, for all the tribes of the Americas. There was a great difference in perception between European and Native cultures. The Natives had a vision of the world as one of mystical empathy and care; the tribal hunters had songs for the whale, even for the harpoon. They asked the whale to offer itself. They saw a world in which everything was, is, alive. This view contained a scientific knowledge that is now called ecology.

The Europeans searched this continent sometimes for wealth, sometimes for their own needs and survival; their own continent was deforested and already contained few wild places. For them, this land was one of surprises, some frightening.

On the West Coast, one white man described the forest and

wilderness of trees: "Their deep and impervious gloom resembles the silence and solitude of death." On the East Coast another described "the dreadful frogs" whose croaking frightened them. It was all-disquieting to them, this world of beauty: "What could they see but a hideous and desolate wilderness, full of wild beasts and wild men?" asks writer Richard Ellis.

This fear was, in part, what allowed for damages to the natural world. On the East Coast gray whales then existed as far north as New York and as far south as Florida. These Atlantic grays were hunted into extinction by whalers who took only the oil from the whales, leaving the carcasses behind. The whalers carried along with them the burden of themselves. Animals became their fears, as did forests, and they endowed them with their own morality and their own emotions. They opened up the New World full of fears, beliefs, and needs; then they closed the world behind them, even kidnapping Native peoples to display in Europe.

The newcomers thought this New World was a place of evil. When they heard the clamor of seabirds from one island in the Atlantic, they believed they were hearing the voice of demons. Looking back, we can see that the European arrival here seemed to be, to bring, a fall, as if the world changed to fit their beliefs. It was, at first contact and in their own words, a paradise. By their own actions, they fell out of this paradise. In their minds and words, they transformed the world into something it should never have become. Not only did they want gold, but they also layered the American continent with a view of the natural world that did not accept that the Earth was alive and that all species were sentient.

We can only imagine the great numbers of whales at this time. They were magnificent, yet called monsters. You can read their history and fate by the names Europeans gave to North American waters: Devastation Bay, Shipwreck Trail, Whale Town, Destruction Beach. They are the names of betrayals.

~

This land had been occupied by a culture for thousands of years before Europeans arrived. European demands for fur and fish in exchange for spoons, metals, buttons and other items helped to break the ecosystem, according to author Ruth Kirk. Many whaling tribes, including the Hesquiat, near Nootka Sound, lost populations to European illnesses. They were even forced to move away from Yuquot, the "place where wind blows from all directions," to a new location. History has been as tumultuous as the winds in this place.

West-Coast whalers, newly come to coastal waters, described the whale simply as "a certain fish of considerable size. Blowing water out of its head." By the mid-1800s Monterey, California, became one of the largest whaling stations, and Trinidad, California, was an island soaked in oil and fat. The bodies of these animals became candles, soap, leather cleaner, corset stays.

~

Among Natives, the marine mammals, including whales, walrus, otters, and seals, were their lifeblood. Now their numbers were depleted.

When the white men came, tragically many Native people ended up participating in the extinction of the animals that they once had depended upon. These people were broken and changed; many lost their spiritual life, their connection with the whale. The relentless slaughter was not limited to gray whales. Manatees were killed on the West Coast, never to be seen again. Bowheads were slaughtered by the thousands.

There were also conflicts between nations that resented others for taking "their" whales. Whaling on the "hell ships" brought out the worst of men. Their lives were dangerous. There are stories of splintered boats and men thrown through the air by whales, the destruction of instruments and equipment.

Mutiny was common. Even Melville never completed a voyage. In his last ship-jumping, he remained in Hawaii.

I have often wondered if the female view might have been different from that of the men, but this is not necessarily the case. On at least one whaling ship, the Addison, *the captain's wife kept a journal.*

> Aug 26, 1858: "Nothing that lives. Our company was beginning to feel rather down again when the cry of white whales resounded from the head. The boats were lowered and about eight o'clock p.m. got a whale alongside. Commenced cutting him in immediately...."

> September 2: "Whales in abundance. It is a grand sight to see them ploughing through the sea, rising to breathe. If they were aware of their strength, how few would be safe. Our boats went off in the morning for whales. About 10 a.m. fastened to one, which knew how to use his flukes very scientifically. At dark they were compelled to cut from him, which they did with a very bad grace."

The distance from the whale, its separateness from the human, in no way resembles the way Natives saw the land, ocean, and seas.

When I look back on history, I see it as a history of weapons. The tribal nations here were helpless before the advanced weapons and against the great numbers of Europeans. The Aleuts fought the Russians and in return all those on Unimak Island were killed by a Russian man, Solovy, and his crew. He'd already been named by the Aleuts, "The Destroyer."

When the white men arrived, there was a deluge of ships. They followed straits and sounds, went across fog-covered water. Soon these places were no longer full of whales. The new methods of whaling used guns, bombs, and lances. The whalers hired Eskimos, and they joined into the cruel occupation of whale killing from ships. Females were lanced as they held their nursing offspring in their flippers, "uttering the most heartrending and piteous cries." One female beached herself, as if the onslaught of whaling had broken her spirit.

Many whale men felt guilty about this butchery. Soon the whales in the north were nearly gone. The bountiful wilderness was lost—on land with the fur trade and forestry, at sea with the seals and whales. Sven Waxell wrote that early on a Native asked of the first Russian traders who invaded their lands near Kamchatka: "If you could enjoy these advantages at home, what made you take so much trouble to come to us? You seem to want several things which we have; we, on the contrary, are satisfied with what we possess, and never come to you to seek anything!" In a short time there was nothing to take and when the Indians were rebellious they were killed.

Behind the Scenes

Peace-seeking dialogues between moderate Makah and conservationists have continued discretely since 1995. And during the past years, both John and Micah McCarty, as well as Alberta Thompson, Dotti Chamblin, and other Makah elders, have journeyed around the world to talk about the tribe's whale hunt.

"It is a different world today than it was yesterday," said one of these conservationists after meeting with a Makah tribal member for the first time. "Oh, if we can just give ourselves more time to get to know each other before our history of war again overwhelms us."

There are some still holding this fragile vision between Native and non-Native peoples, between humans and whales. For we have imagined it together.

That summer night of 1998, we took our leave of Micah

McCarty, but we pledged to meet again. As we left Micah's kitchen table, Toni Frohoff said quietly, "I don't know what to do now, except keep protecting the whale. But now I also hold dear one who holds the harpoon."

We did not know, as we left the reservation to drive the dangerously winding roads along the Pacific Coast, that in the next two weeks Micah would also take his own leave. In a surprising decision, Micah McCarty left his position on the Makah whaling crew. Announcing publicly that he was returning to college in nearby Bellingham, Micah McCarty and his wife, pregnant with their first child, left the reservation. Micah's unexpected departure came in the midst of much turmoil on the reservation.

Some of the Makah were unhappy with the whaling crew. One member had been seen breaking the strict purification and cleansing regimen of a Makah whaler by drinking beer in the midst of preparations for the sacred hunt. Rumors spread about drug abuse among some whaling crewmembers. Mixing substance abuse and antitank weaponry made a dangerous cocktail, especially when the world was watching.

Then, in a very public and embarrassing incident, the whaling canoe capsized into cold waters during a practice run. The media also recorded a support motorboat towing the whaling canoe around Neah Bay. The public reacted with strong disapproval to the sight of the crew relying upon modern equipment, rather than traditional skills.

And then there was Micah's disturbing dream: He and his whaling crew harpoon a gray whale only to find the creature dragging their canoe over the rocky coastline and then diving down, down into the ocean and through a huge rock formation. The wounded whale escapes; but the men cannot follow or find their way through the cold, undersea rocks.

When Micah told Linda and me his dream at a whale conference, he was very concerned about the spiritual unison of his whaling crew. Many whalers have reported that when a gray whale is struck with a harpoon, the whale lets out a banshee's wail that will haunt a man for the rest of his life. The only balance to such a cry from a wild animal would be the firm conviction of a whole tribe's spiritual and physical subsistence.

Left Out of the Equation

When Americans took over the whaling trade, the Natives who had hunted for them in the past were left out of the equation, or worse, treated badly, the leaders even humiliated. In this, there was the tragic breaking of harmony and relationship. One company, the Davenport (later called Portuguese), numbered its annual income from whaling in the millions. For tribes, the world was destroyed.

Whaler Robert Gray burned one village, jealous he could not purchase it; it was a meaningful place, beautiful and elegant with its carvings. Travelers always remarked upon carvings along the Northwest Coast, yet the Americans thought of the people and their beauty as "primitive," not as radiant, primal, and spiritual. In this one village, Gray wrote, the people entered houses through the mouths of animals. It must have been a beautiful sight and surely these took lifetimes of building, carving, and care. The Native canoes were also ruined so tribes could not go into the sea and feed themselves. Worse, the tribes were turned against one another, making and breaking alliances. Then came sawmills.

And then it seemed the land, the ocean, everything was emptied. There was the breaking down of people, and the hatred by the whites. There were epidemics, alcohol, and settlers. The Eskimos, and rightly so, burned one ship in 1872; another, which contained 9,000 gallons of sperm whale oil, was lost at St. Lawrence Island. Another carrying whale meat for fox farmers crashed when its own whaling gun exploded in 1920.

Called to Compassion

Neah Bay, Washington, Spring 1999

Chuck Owens is a lifelong fisherman, and his wife, Margaret, is a potter, most of whose work is devoted to making ceramic gray whales and tiles. The couple lives very modestly in Joyce, Washington, right next door to the Makah reservation. A bumper sticker on their battered pickup reads, "The best things in life are not things." Margaret Owens is 51 and her husband is 48; all but one of their three children had left home when the couple, faced with the neighboring Makah tribe's whale hunt, found themselves called to compassion as unlikely activists. "We're speaking for the whales," Chuck Owens said.

A towering mountain of a man in Oshkosh overalls and graying beard, Chuck said, "The last thing we want is a commercial whaling operation off our coasts. That tragedy belongs to the past."

So the couple formed the small, grassroots group Peninsula Citizens for the Protection of Whales. They took to the roads, not the waters, in their effort to protect the gray whales from the Makah hunt. They called it a "land-based protest."

With an active membership of just over 20 people, "mostly mothers and grandmothers," Chuck said with a grin, the PCPW sponsored rallies, bake sales, whale adoptions, and roadblocks to bring public attention to the plight of the passing gray whales targeted by the Makah hunt.

Owens noted that in 1994 the Makah tribe quietly returned to seal hunting and had stated their intention to also return to hunting the sea otter. Chuck and Margaret Owens have found themselves, like the anti-hunt Makah elders, trying to preserve a culture of respect for gray whales, a culture that their friends in the Makah tribe very much support privately, but are afraid to voice publicly. "Sometimes I think we're also speaking for all those Makah friends who cannot risk their Council's wrath by sharing with the outside world. So they tell *us*," Margaret Owens said.

Chuck Owens has fished and worked alongside Native peoples from Alaska to Washington and had a long history of friendship with many Makah. Two years after the first hunt, he was no longer allowed on the reservation. "We still have many friends in Neah Bay," he said sadly, "but we can't go visit them anymore. Our presence would incriminate them. But many Makah are not for this hunt and have been helping us quietly."

KATY PENLAND, AMERICAN CETACEAN SOCIETY

Close to shore, a gray whale "spy-hops" near Palos Verdes, California.

"White Beak" provides rare documentation of a gray whale calf feeding on eelgrass in San Ignacio Lagoon in Baja, Mexico.

L.A. HENDERSON

JARED BLUMENFELD/IFAW

ABOVE: Makah elder Alberta Thompson encounters a gray whale mother in San Ignacio Lagoon in Baja, Mexico.

MARGARET AND CHUCK OWENS, PENINSULA CITIZENS FOR THE PROTECTION OF WHALES

LEFT: Signs from whale advocates line the road to the Makah reservation during the spring hunt in 1999.

PAUL RADCLIFFE

A gray whale calf breeches and frolics in a winter Baja birthing lagoon. At about three months, calves begin their long spring migration north to summer Arctic feeding grounds.

ANONYMOUS

The Makah hunt-support motorboat, the *Heidi*, tows the harpooned whale back to Neah Bay during the spring hunt in 1999.

ERROL POVAH

Makah Whaling Captain Wayne Johnson watches as the carcass of the harpooned whale is butchered.

L.A. HENDERSON

First Beach and A-ka-lat, "Top of the Rock," provide the sacred burial ground for the Quileute tribe's ancestors. The Quileute tribe is planning a "Welcoming the Whale Festival" here to celebrate the gray whale's passing each spring in La Push, Washington.

L.A. HENDERSON

RIGHT: Fred Woodruff, his family, and other tribal paddlers have journeyed over 4,000 miles since 1989 in this Quileute 33-foot cedar canoe, the *Os\chuck\a\bick*.

BILL HESS © 2000

Three gray whales were trapped in 1988 near the Inupiat village of Pt. Barrow, Alaska. Two of them were rescued by a coalition of Eskimos, the U.S. military, and a Russian icebreaker.

Monterey Bay, California, is known for its deep submarine canyon and orca attacks of vulnerable gray whale calves. This is the first photo documentation of such predation from April 1998.

SUE FLOOD, BBC NATURAL HISTORY UNIT

PHILLIP COLLA

Off Big Sur, California, a newborn gray whale calf, estimated at 6 hours old, playfully engages divers with its umbilical cord still attached, while its mother floats nearby.

The Owenses talked about their fears. "The money and energy spent on this hunt, when there are so many other pressing priorities, is tearing the Makah tribe apart," Chuck Owens said with a shake of his head. "The first whale killed will lead to 'cultural' whale hunts." He cited Japan and other whaling nations' demands for approval of their own cultural whaling plans.

"Thousands of other whales will be killed," Owens said. "And it all begins here. We feel we really have the pulse of the people on all sides. And the work we are doing is for the good of our community *and* for the whales."

Protectors of the Whale

During spring 1999 Makah whale hunt, activists calling themselves "protectors of the whales," not "protesters," roused themselves to climb aboard their borrowed boats and patrol the waters around Neah Bay.

"The newspapers call us *anti-this* and *anti-that*," one of the young men told me. "We're not against things, we're *for* something...we're for serving and saving life. The life of a gray whale is just as important as any other life."

That protecting the whale had come down to so few—by some estimates, fewer than 50 protesters—was a sign of how very divisive and bewildering this tribal hunt truly was. The international media camped at Neah Bay by the hundreds, awaiting what some feared would be a 20th-century replay of conservation cowboys and Indians on the high seas. All eyes were on the waters around Neah Bay.

While the Makah had the weight of the federal government's support behind them, the activists had few resources and even fewer boats. They subsisted on donations called in to a local motel near their camp. Acting without the support of major environmental organizations, the young activists often felt that theirs was an impossible job. Protecting the whales was an awesome responsibility to be placed on such young and inexperienced shoulders. Their daily regimen was grueling.

"What the public doesn't realize," explained activist Will Anderson, "is that these people, the youngest and, in many ways, the cutting edge of the newest environmental ethic, went out every morning about 3 a.m. to make that one-hour trip to waters outside Neah Bay in the total darkness and cold—just so they could be there if the whales were in danger. Every day, no matter whether the Makah were officially hunting or not, the activists were standing guard as best they could."

In those volatile spring days during the Makah hunt, activists were also doing everything they could think of to gather public support and awareness. They sent a homemade card to the Quileute reservation: THANK YOU FOR LETTING THE GRAY WHALE NATION PASS IN PEACE.

In May 1999 the Makah whaling crew set out again to hunt. As the whaling canoe slipped into the cold waters off Neah Bay, the Coast Guard was rigidly enforcing an exclusionary zone and arresting activists and their craft. Native canoes were paddling into Neah Bay from the neighboring Quileute, Puyallup, Tulalip, and Hoh tribes to show support for treaty rights. Reporters followed each other around, desperate for a story. The television broadcasters covered the issue like a sporting event with live helicopter cameras and instant replays of harpoon throws.

Contradictions abounded. At the same time the U.S. Coast Guard was arresting whale activists trying to stop the hunt, they were also charging activists, under the Marine Mammal Protection Act, with harassment of gray whales. The gray whale itself was often lost in the media circus.

One activist, Captain Paul Watson of the Sea Shepherd Conservation Society, told the press, "I don't know what kind of a crazy farm I've just run into here but these guys want to blow a whale apart with a .50-caliber machine gun and everybody's concerned that we might get too close to a whale."

I will not ever forget that misting morning, May 17, 1999, waking too early. At 6:30, I sat straight up in bed and instinctively reached for the remote control. At the moment my tiny television blinked on, the Makah crew was closing in on a small gray whale. At 6:49 a.m., a Makah man stood up, balancing in the canoe, harpoon raised as the whale glided alongside.

It was all so familiar to me—the curious whale, perhaps a Friendly, nearing a small boat, rolling sideways to gaze up at the excited people. Making eye contact. So intimate, trusting.

But this was the Makah tribal canoe, the *Hummingbird*. A Coast Guard boat patrolled the federal "no-entry zone" in a huge circle around the whale and hunting canoe, keeping out a flotilla of smaller protest boats.

With an entire world watching live, the juvenile gray turned on her side sliding right up to the canoe. The Makah scrambled for their weapons. The gray came closer, right alongside the canoe. Close enough to touch.

And for the first time in 70 years those dark blue waters ran bright red with the blood of a harpooned gray whale. Millions watched the kill live on television as the wounded whale struggled, twisting and zigzagging, pulling the Makah hunters a short distance, three harpoons lodged in her 30-foot body. Bloodied waters swelled outward from the dying whale in crimson waves. Three times the hunter shot the young gray, one bullet through the brain in a mortal explosion. At this point many of the TV cameras cut away, deeming the kill too grisly for viewers. But in real life, it continued.

After several minutes of struggle the juvenile gray whale submerged—not of her own accord, for she was dead. As her open mouth took in seawater, she sank down and down. No one among the hunters had the equipment or the skill to dive into the water to sew up the mouth of the gray so the huge body would float, as their Makah ancestors would have done. So the gray sank to the ocean floor where she would ordinarily have stayed, resting like a shipwreck.

What should have involved several hours and a triumphant tow to the nearby shore turned into an all-day ordeal. The Coast Guard assisted the Makah by retrieving the sunken whale. Using an air pump, they inflated the dead whale enough to coax the carcass upward with an onboard winch. With assistance from their motorboat, *Heidi*, towing the dead whale toward shore, the five canoes, including the *Hummingbird*, floated the carcass to the beach at Neah Bay. By early evening, the world watched on the international news.

A hunter jumped onto the back of the whale with a rattle and sang a song. Other young men clambered atop the broad back of the gray whale and raised their fists in the air, exchanging triumphant high-fives. Some of the celebrants did back flips off the whale. News cameras rolled while some Makah claimed victory and triumph.

"It was easy," said whaling crewmember Darrell Markishtum. "The whale gave us up his life freely. He didn't fight."

Others in the tribe stood off in the distance. Elder Alberta Thompson wept.

Helplessly, I placed a hand on my television screen as if to comfort the dying whale as the news programs played and replayed the edited tape of only the first harpoon throw, the whale's futile bolt from the boat, no blood. I sat stunned on the floor watching it all, as did millions throughout the world.

After the television cameras had stopped rolling, most of the Makah hunters went home, leaving few, if any, to complete the ancient ceremony of singing all night to the sacrificed whale. That night a man the Makah had hired from a subsistence whale-hunting Inuit tribe patiently flensed the whale. He labored long into the night with little company except an official representative from National Marine Fisheries Service (NMFS), who had to oversee the whale hunt to be sure it was done within the rules agreed upon between Makah and the U.S. government. To many Makah this abandonment of sacred tradition seemed a grave dishonor to the spirit of the young gray.

So that long, lonely night, the Inuit man butchered the whale, sawing through surprisingly thin layers of blubber. At one point after midnight, the Native carver called out for help from the small crowd gathered around the gray whale's carcass, "Are there any Makah here?" he asked. According to reporter Robert Sullivan, "no one responded. One of the reporters decided to help the whalers and so waded into the bloody midnight water and assisted in the unraveling of the whale's guts."

As one Makah elder said, "When the man kills the whale, the only part he takes is the saddle or the dorsal fin. He takes that to a smokehouse for four days. He meditates and prays. On the fourth day, the spirit of the whale leaves. Well, these kids took the dorsal fin and put it in the backyard and left it there."

In a video taken by undercover anti-whaling activist Erin O'Connell to document that night, there were no Makah whalers present at the butchering. A National Marine Fisheries officer overseeing the Makah hunt was so frustrated with the lack of ceremonial follow-up on the part of the Makah whalers, he told them, "We shouldn't have to be doing this for you guys!"

After the hunt, the backlash against the Makah was intense and immediate. Harsh criticisms were leveled against the Makah and the U.S. government from all over the world. Angry letters filled page after page of the *Seattle Times*, with comments running ten to one against the hunt.

The Tribal Council hosted a huge potlatch and invited other tribes to Neah Bay to celebrate their success. A feast highlighted songs, ceremonies, dances, and speeches supporting the exercise of treaty rights. It was the first time many had tasted whale meat.

Chuck Owens noted that there was no known photo identification of this three-year-old gray whale. "But she was seen in the company of nearby resident whales," he said. His letter to the editor in the *Peninsula Daily News* strongly refuted the NMFS statement that the whale killed was unlikely to have been a resident whale because "it was taken several miles off shore, traveling north," and thus probably a migrating gray. "According to the NMFS's own official observer's report, that whale was never more than a mile and a half offshore and traveling *south*, not north like a migrating whale."

Owens added that on May 20, three days after the successful hunt, "marine biologists with Cascadia Research photo-identified six resident whales in the area where the whale was killed."

The small, seemingly faraway Makah tribe is no longer isolated from a global perspective; it must now weigh its decision for future generations. The prestige and spiritual respect given the Makah whaling ancestors in another time may not be granted the next generation of Makah in today's world as they return to whaling.

It was Makah elder Alberta Thompson who gave a name to the young gray whale taken by her tribe. *Yabis*, she called the whale. In the Makah language the word means "beloved."

Reckoning With the Spirit of the Whale

"They haven't reckoned with the spirit of the whale," Alberta Thompson says. And this is true in more ways than one. There are consequences to whale hunting, not only in the fight of a whale sometimes known to turn on its hunters, but in the fact that in the old days the relationship between the people and the whales was the significant factor to every whale hunt.

As Indians, we must speak out for both the old people and the old ways. In the traditional and historic past, we recognized the sovereignty of other species, animal and plant. We held treaties with the animals, treaties shaped by mutual respect and knowledge of the complex workings of the world, and these were laws the legal system will never come close to.

In their location at the end of the continent, a people are trying to lay claim to an older world and its complex of ceremony, but which people? It may very well be the silenced older women.

The future generations will look back to these times and will see these women as courageous as our leaders were during the treaty-making times when they spoke to their tribe and to the outside world.

There is another possible outcome to this story. If the Makah choose not to continue whaling, it could truly make a statement about how strong a culture can be, as it looks to other means for the true and deep wellspring of a culture, of a people, one that holds to a reverence for life.

They will set an example for others by which part of the culture they decide to cultivate. And for the children at Makah, what better example than seeing their own people take the side of life, as part of the sacred. That might very well restore tradition until the whale and the people reestablish a relationship of offering and receiving from one another. The way it used to be. The heart of the hunter has to care.

Between a Harpoon and a Whale

The Makah spring hunt of 2000 began with renewed intensity on both sides. But in this hunt, it would be photos of a young protester in an almost fatal collision that would make international news.

The Makah had been hunting in family canoes since April 17. On April 20 a Makah whaler threw a harpoon at a young whale. At that moment, a protester on a Jet Ski sped between the hunters and the harpooned whale. Perhaps the young whale was unharmed; perhaps it was already wounded. No one knows. What happened next would be played and replayed on television for months afterward: A Coast Guard boat ran right over the protester. Minutes passed before she bobbed to the surface, injured but alive.

When I interviewed the slightly built protester, 24-year-old Erin Abbot, she was keenly aware of her near miss with death. Her ribcage was still wrapped in Ace bandages. As she told me her story, her voice was very soft.

"On a Jet Ski it's very hard to see the Makah canoe over all the ocean swells," Erin began. "I saw a Makah man stand up and throw a harpoon. Then I saw a blow and water spewing up. So I just gunned it and got in between where I thought the whale was and their canoe." She paused, her voice shaken, "And the Makah were trying to hit me with their paddles, but I was a good six or seven feet away. I veered to the left to get away from them and that's where on the TV video you see my Jet Ski spray going over the back of their canoe. The press makes it out that I was being violent by trying to swamp their canoe, but I was just trying to avoid being struck by their paddles.

"I turned away from their canoe and that's the first time I saw that 21-foot Coast Guard Zodiac bearing down on me. In seconds, they ran right over me. That aluminum bottom felt like cement hitting my right side. Suddenly I'm underwater and right under the Coast Guard boat. I knew the twin-engine propellers would kill me so I dove way down. If the Coast Guard had knocked me unconscious, I would have died from the propellers. And if my Jet Ski had been just one inch farther back, they would have ridden over my head and killed me as well."

Erin was at last hauled up from the water onto the Zodiac that had just hit her. She was suffering from hypothermia and what x-rays would reveal as a shoulder broken in half, bones overlapping. Her breathing was labored because of four broken ribs. Airlifted to an ambulance in Neah Bay, Erin pointed out that the Neah Bay reservation clinic doctors were very fair.

"They did a good job and were very kind to me," she remarked. "As they put the neck brace on me, I told them, 'I'm not against the Makah. It's just that the whales don't deserve to die.'"

Erin Abbot was held under guard in the Olympic Medical Center in nearby Port Angeles for five days. When she was released, she was taken to a Tacoma court in handcuffs and shackles. The judge released her on her own recognizance and at a later hearing she was ordered to do 120 hours of community service with a nonprofit, non-animal-rights organization. She was also forbidden to enter the Makah Reservation and was put on probation for one year.

Letters and campaigns on the Internet during the highly contested presidential elections in 2000 were full of fierce debate about the future of Native and non-Native whaling. Ojibway-Cherokee artist and activist Linda G. Fisher wrote, "True Indian animal activists are few and far between, but we are out there." Noting that her uncle had stood beside Indian activist Leonard Pelletier when he was arrested, that "my mother is a respected elder of our clan, who also experienced the pain and heartache of survival in a white man's world," Fisher concluded, "I am not speaking as an outsider.... As Indian people we must fight our war for equal rights, regain our land, and stop corporate pollution of our reservation land without using and hurting the fragile creatures on this Earth that we as a people used to be so connected with."

Sea Change

In June 2000, the Makah spring hunt was unexpectedly put on hold by a San Francisco court of appeals. Judge Stephen Trott questioned the objectivity of original environmental assessments by the government, noting that a policy choice already made to allow the Makah to hunt might have "slanted" its analysis. According to the *Seattle Times*, the federal judges not only called for a new environmental assessment, but for one "done under circumstances that ensure an objective evaluation free of the previous taint." By a vote of two to one, the court reversed an earlier ruling that allowed the Makah to resume whaling and ordered a new study of the environmental risks.

When that final Environmental Impact Statement (EIS) was made public in July 2001, it shocked many environmentalists, not only because it did not protect "resident whales" in

Neah Bay, but because it in fact expanded the hunting waters from open ocean to inland areas of the Strait of Juan de Fuca between Washington State and Canada. The new EIS also opens the hunting season to any time of year, instead of restricting the hunt to spring and fall migrations. Though animal welfare groups, such as the Fund for Animals and the Humane Society of the United States, have again filed an injunction challenging this hunt and asking for another environmental assessment, the Makah hunt seems to be legally back on track.

But since the successful hunt of May, 1999, there has been what Lynda V. Mapes—longtime *Seattle Times* reporter on the Makah hunt—calls a "sea change" on the part of the Makah tribe toward whaling. In the spring of 2002, a new Makah Tribal Council, while continuing to assert its treaty right to go whaling, is now deeply concerned with a 50 percent unemployment rate and is arguing that the tribe's "other needs are more pressing." The focus now is on "economic development and jobs. Public safety, drug enforcement, programs for youth, and even tourism all are on the council's agenda."

In 2002 the federal government is no longer so heavily funding the hunt. Since 1996 the government has contributed over $360,000 to support Makah whaling. These monies paid for tribal delegations to the IWC, research to assure a humane whale-killing method, a tribal biologist, and other support during the 1999 hunt. (The National Marine Fisheries Service makes clear that none of that money was spent to actually kill a whale.) In May 2002, after bitter debate and international deal-making, the IWC at first revoked and then reaffirmed the Makah quota of 20 gray whales to be "harvested" through 2008.

Reacting with disappointment to the changes in both federal and Tribal Council policy, Wayne Johnson, captain of the whaling crew that killed the juvenile gray whale in 1999, said he may never go whaling again. "It hurt me so much, I sometimes wish I had never gotten involved."

Whaling is now considered more of a private, family affair by many Makah. Although no whales have been killed since the single one in 1999, as many as three families still express active interest in another hunt. "Whaling will continue forever," says Greig Arnold, Makah whaler.

One man close to tribal politics adds, "The Makah are patient people. And it will depend upon their leaders as to whether they will continue to actually kill whales. Perhaps they've proved their point and can move on now to other successes that includes a better economic future for all tribal members."

Will this sea change among the Makah themselves also invite a new beginning between tribal members and non-Natives? Or will the Makah, with the help of other whaling nations, help return us to a world of whaling? Will the continuing, quiet coalitions of moderate Makah and conservationists at last make a bridge strong enough to carry all cultures, human and animal?

PART FOUR

Eye to Eye: Honoring the Whale

Ozette, Washington, to Vancouver Island, British Columbia

1995 to 2002

Other Visions, Other Futures

The Makah are not the only Northwest tribe to have traditionally hunted and intermingled their ceremonies with the great gray whale. All up and down the West Coast, from Chumash hunters in California to Arctic Eskimo whalers, indigenous peoples have depended upon the grays for thousands of years, not only for their physical survival, but also for their spiritual subsistence. Stories of whales and Native peoples tell of this ancient bond—in art and dance, in hunting tools, and oral traditions that, as we enter the 21st century, are still vital.

In following the grays and the people who find meaning and interdependence with them, we also hoped to discover other indigenous visions that, while flowing from the same past of whale hunting, are now adapting these sacred traditions for a new and environmentally more fragile world.

Beginning with Ozette—one of the most famous of all Native archeological finds on the West Coast—visiting the neighboring Quileute tribe who are turning their whaling canoes into whale-watching canoes, and then profiling the vibrant First Nations people of British Columbia, we continue our journey northward. The journey shows us both the splendors of the past and the possibilities of the future.

Deluge: Ozette

In nearly all cultures across the world are stories of the great flood. In some stories it rains for 40 days and nights. In others a wave of water submerges an island or consumes part of a mainland. Land is sighted and then lost. The cataclysmic flood is a story held by all the indigenous American Northwest Coast peoples.

In the oral traditions of the peoples along the West Coast, especially those near La Push, Washington, and Neah Bay, and further north with the Nuu-Chah-Nulth people in British Columbia, are stories of earthquakes and tidal waves that washed away towns and beaches, waves of water that came over the houses. There are accounts where the water rushed so violently across land that destructive fragments of trees whirled over the swiftly moving water-covered world.

In one account, the water, "on its rise became very warm, and as it came up to the houses, those who had canoes put their effects *into them, and floated off with the current which set very strongly to the north." Some drifted one way, some another; and when the waters assumed their accustomed level, a portion of the tribe found themselves beyond what is now called "Nootka," near Vancouver Island, where their descendants yet reside.*

Stories are also told of how the mud swept into the villages and knocked down houses, burying them so deeply that people and possessions were lost forever. Mudslides covered entire villages. This was what happened at the ancient Makah village of Ozette, 30 miles from Neah Bay, Washington, where Cape Flattery marks the westernmost point of the lower 48 states. Buried perhaps 500 years ago, Ozette began to reappear around 1970 when a storm uncovered its remains.

The mud had sealed and closed the world so completely, so quickly, that the alder tree branches were still green when this village was first found. They aged only afterward, when exposed to air. Spring pollen was found on the ground. This was a world caught in its own midst. It was a world held still..

The appearance of Ozette reminds me of a story from the Northwest Coast I once heard, about a mythic Kwati box containing daylight that had been stolen. Perhaps this newly found place may be that box, found, opened. The discovery of the old village once covered in mud is an archeological find equal, one scholar writes, to Pompeii. Certainly it reveals much about the people both before and after European contact. It is, truly, a revelation. It is a record of history

and a reflection and affirmation of what elders have said. It validates stories of the Quileute and Chemakum, among others who are from geologically unstable places along the coast. Tidal waves, mudslides, and floods affected all the tribes in this still-teetering world. It is a place of fault lines, geological, historical, and human, all as torrential as the floods that covered the village.

With the revelation of this village named Ozette, memories also surfaced, as if they too had gone down beneath time and history, one of the many mysteries we still do not know about this Earth.

When the village of Ozette revealed what had been lost to history, it created change for many of the peoples along the coast. It seems for many, the Salish, the Makah, the Quileute, the Kwakiutl, and others, including the cousins at Tofino, British Columbia, that this newly uncovered village held the saved pieces of a broken world which might now be mended, Whether the people heal from their histories or not, only the future can tell. If the spiritual life from ancestors can be reclaimed lovingly from the past, into now and the future, this discovery at the village called Ozette will make for a fine new way of an older life.

The Approach

We see the black rock in the silver ocean. Seal Rock. Its name tells us what used to be there. It is smooth on top as though a glacier once passed over it. There is a hint of sun and a mist from the sea. Together it is neither cold nor warm. These days, this ambiguity says everything about the place, the geological history, and the people who were here or passing through. As with all the villages along the coast, the people here were deluged not only by tidal waves, but by a breaking history. This history also arrived from the ocean and washed over the people of this coastal region.

This area, jutting out into the ocean, was the perfect location for a trade center. In 1788 the Europeans arrived in boats so large the newcomers were called "The House on the Water People." They first began a trade in furs, with a preference for otter fur. Later, they would be the ones who nearly extinguished the gray whales, hiring

Native men. As a trade center, the place was overtaken by many, a flood of Europeans, the Spanish, British, even the Chinese. Then there was the smallpox epidemic in 1852. Rapid cultural change and epidemics created massive losses of life; as much as 90 percent of the Native people died soon after first contact.

By 1877 Neah Bay was a reservation trading post, then an Indian Agency was set up, and then the children were forced to go to mission schools. So began the typical attempts, the designed process of assimilation. Because of their desirable location, these villages had stronger forces arrayed against their traditions than did tribes in other locations.

Now, as in other places, few remember how to behave with the Earth. With the discovery of Ozette, this no longer seems beyond the memory of the people.

We find it true that whaling was part of a spiritual life and a noble calling. The people on this coast used cedar canoes while those up north used skin boats. Still, the ceremonies were similar, a whale approached with awe, reverence.

There are sediments and layers of living, some called prehistoric, meaning before European recorded history. In those places we are able to find the deep history and meaning, the memory of the people before contact. Now, found history is more than "only" stories.

As in the past when the sun dropped across the world and the people watched it, the sun shines again on the village and we are fortunate enough to see some of the things that were revealed at Ozette.

At Ozette, scholars, Native as well as others, have concluded that the Quileute were the earliest to inhabit this region, long over four thousand years ago. They decided this because the place names and words for features of landscape are, were, in the Quileute language. The Makah arrived later. Yet there are many similarities between the two, both having once whaled and killed gray whales.

Quileute Chief Howeattle described the whale hunts he'd experienced to writers Rowena and Gordon Alcorn. The first hunt occurred when Howeattle was only 14 years old, in 1899. This was after European contact, so the hunts were not the elaborate ceremonial occasions of just a few years before in which a whaler prepared himself carefully, as did his wife. Howeattle says he heard the cry, "Whales spouting! Whales spouting!" He was excited and soon they were past a line of breakers and they approached the whales. That night they made a fire and camped on the side of an island hearing the sprays and breaths of the whales. Before dawn they were on their way, carefully because there were so many dangers; the waters, the whales, the delicate balance of the canoe and its men. Soon, again, they began to hear the sound of whales blowing. They saw another tribe coming from the Hoh river, Howeattle told the Gordons. The others harpooned a whale and towed it back to the mainland. Soon, more Quileutes arrived to help and finally one old man harpooned the whale. Howeattle says of himself, "I was so scared that I started paddling the wrong way." The whale was the biggest thing he'd ever seen.

Howeattle's job was to attach the sealskin buoys to the whale to keep it afloat. Four years later he went out again; this time he threw the harpoon and soon, he said, the sea was red with blood.

"It took two men to pull out the harpoon. I said to myself, no more whales for Charlie."

Most tribes in this area didn't whale. Nevertheless, they showed the whale great respect. With the neighboring Haida, their word for power was, is, the same as the word for whale.

For those who did whale, there were ceremonies, ritual bathing, fasts, and prayers before hunting started. They began an "entreaty," a reckoning with the spirit of the whale.

It seems so present when you see that the people of Ozette had cleared a dragway, still there, to haul in the sea mammals. As for the whale, after keeping it afloat, after sewing its mouth shut so it wouldn't fill itself with seawater, the people used ropes to haul it away from the water home while the men sang, waiting for the help of the tide. All the while, the people lived in silence and respect for the taking, the bringing in of the enormous life. Everyone, even those on land, moved gently, talked quietly, and with love in their hearts. That's what is said.

Bones

The gray whales passed Ozette on their long migrations through the coastal waters. Evidence reveals that the Ozette people ate mostly smaller sea mammals, those being easier to reach and not so dangerous to hunt. The bones found at Ozette are not generally identified as gray whale bones, and it appears that humpbacks may have been killed in equal numbers. Nevertheless, the people in the region were whalers; but they may also have been like the Chumash and other tribes, consuming the many whales that beached themselves, injured by killer whales and the once-abundant swordfish.

Items from both the Quileute and the Makah can be seen at the Neah Bay Museum. And Ozette is not the only old village found along the coast; other sites have similar finds, including places in California, Canada, and Alaska. But the museum at Neah Bay now has the finest public collection, an impressive display that takes a visitor back in time.

In the Neah Bay Museum, the bones of the whale seem dead center, as if the whale is as central to the museum as it once was to the people. The bones have a magnetic quality. Sea mammal bones are light and porous. They carry sound, including the human songs and prayers as the people travel in their canoes singing to the whale. They carry sound through water in a language deep and difficult for most humans to hear. .

Interestingly, many of the whale-related items at Ozette are wooden carvings of the fins of killer whales, designed and embedded with sea otter teeth. The people probably admired this great hunter who left many injured gray whales to wash ashore. There is on display only one stone sculpture that is possibly a gray whale, small in size, dark and mottled. Having no large fin on its back, we can only assume it is a gray whale. It has an indented neck and folds similar to those in the throat of the grays. This opens a question of why depictions of the gray whale are missing from the art.

British Captain James Cook, in one of his many voyages, commented that along the Northwest Coast there was nothing without the figure of an animal upon it. The art, the items even for everyday use were the elegant work of people closely in touch with animals. There were planks, Cook noted, carved with wolves, birds, and owls. There was also a plank with the carving of a killer whale.

It was said that one of Cook's men, in order to sketch some of the artifacts of the tribes, traded his brass buttons.

One of the earliest items found at Ozette was the finely carved handle of a whale-bone club. It was similar to finds in other explorations at Nuu-Chah-Nulth, at Hesquiat Village, Toquaht, Kupti, and elsewhere in Nootka Sound. This bone was used as a weapon against smaller sea animals. Once harpooned and pulled ashore, these animals were clubbed.

Archeologists also found carved harpoon valves, as well as sealskin floats, which kept the whale from sinking. In a hunt there would be perhaps 14 of these floats used. Sealskins also contained water to drink, for the whalers sometimes were pulled far out to sea, having to paddle back, towing their whale, laboring for four or five days. There are accounts of how easily a whale could not only tow a canoe far out to sea, but could swamp a canoe or break it apart with a single slap of its tail.

More than just exhibits, the objects held within the Neah Bay Museum show how art and spirit meet. You can see thousands of years of respect, not only in the made things, but for all the world around the people. The hunters hunted out of deep need, hunger, and held within themselves as much love for the life of the smaller sea mammals they ate as they had for the gray whale. And they paid tribute to their way of life; religion here, also, was a way of daily being, beauty a constant.

This was a world where art and beauty met with function.

Truly the artifacts in the museum are beautiful. Perhaps beauty is one of the qualities that added spiritual strength to the whalers, those small yet strong men who had so carefully to hunt and kill an enormous and revered whale. Risking their own lives, they sought out a sacred whale with great religious approach and care for their own survival as they were wrapped into the canoe with cedar cloth.

Reflections

Early visitors reported that the Native peoples in this region used mica flakes on their skin. Captain Cook called it "shining sand." I have wondered if that was not merely ornamental adornment, to shine at night by the firelight, as the Europeans thought, but also for the coastal peoples to align themselves with the shining world about them, to keep the human body as part of Earth, at one with the water and the whale. Covered with light this way, they were as brilliant as the whale rising in water, brilliant as even the water itself. It would be a way to connect a human body with all the rest.

By Cook's description, the people lived in harmony. The possessions found at the village of Ozette confirm this. Beautifully carved boxes and bowls were found. Household items included finely woven baskets made of the thinnest of grasses, the work of accomplished weavers. There is a blanket made of cattail fluff and the seeds of fireweed. A blanket made of dog wool and mountain goat wool. Ropes made of spruce twigs. The people knew the world around them, as did other tribes. From cattails to whale bones, they used everything in what we now call an ecosystem.

Beneath the earth's surface at Ozette, thousands of years old, a house was found, with all its implements inside it, bowls, elk antler wedges, even wooden needles and pieces of string. In one wall a whale-bone wedge was used to seal a hole, to fill a crack to keep wind and cold from passing through. So small a thing, so large a feel for life.

And at this place, or looking at the collection in the museum, you think about the dreams of all the Native past, of people—a humble species—seeing the rise of a whale, the knuckles of its back, the tail slap which could break or flood a canoe in a twist of the whale's pain, and you marvel at how the old ones managed to survive.

Ozette, the village, is a place where the dead are not dead.

The museum is a beginning to a world once thought ended. It is an entrance into the past of a people. Going in, a person passes beneath a carved eagle. "No man would set out to hunt or fish *without first attuning both himself and his gear into harmony with the supernatural," wrote one observer. "He knew that all living things have their own particular preferences. Pleasing the spirits was a necessary exchange for the human right to kill an animal or take a strip of cedar bark, or any other resource. Animal People had specific expectations of how they should be handled by humans."*

Two photographs by Samuel G. Morse stand out in my mind. The first is that of a man whose face shows both fear and concentration. Is he a slave from a tribe that kept slaves? If so, would he be alone? Is he a sealer, a whaler, a hunter? The photographs seem to ask questions of their own. From a Native perspective, I wonder, is this posed? Is the man as frightened as he looks? There is no way of knowing.

Then there is the photograph of a man well dressed by American standards at the time. He is named Young Doctor, formerly a medicine man. He was a fine carver of thunderbirds with outstretched wings, and he was said to have animal spirit helpers. No one ever matched his skill at carving canoes, they say, and he carved long into old age.

Carving

A whaling canoe is carved from a single log. The carver begins his work where the tree was felled. The bow and stern are attached after completion, after moving the canoe away from where the tree fell. The outside is black as ebony. The thin narrow paddles, small and perfect, are made for something beyond just speed, for smoothness of motion. They make for quick movement. Like all the rest, they are made of cedar bark.

This canoe in the museum, newly made, replicates the whaling canoes that were so important to the people. The outside is burned until smooth so as not to bubble and make a sound that a whale could hear. The canoes had to be not only works of art but of genius and knowledge of the sea. The hands of great carvers made them, carvers who knew the dangerous waves of the sea and how to make a canoe that could maintain its balance in the worst of situations, exactitude between life and death.

Words

The words for objects and artifacts are important. The names for items from Ozette reveal that this language is not, like English, a language of nouns. It is a language of movement and "flow" like the water around the people. The cedar cloth that goes into the canoe around the man is called "fits in tightly." The word for a paddle is "the device for going along." What the language says about the people is that they are like the sea around them. The visions of cultures are revealed in their languages, too often unremembered now. The world of Ozette itself, the culture, is unlike the reality of a stilled, fixed place; in it you can see motion, sea change, shifting winds, and eddies of history. It is also the whale's world, one of constant movement.

In this place, at the end of land, Brenda and I talk about the whale-bone wedge in the house. It is such a small thing and yet it is something that can be understood, known. Outside, a trail of light across the water is like a path to be taken. The whale spirit gave the songs to the people. We think of that now, listening to the breeze coming in from water.

The Whale Who Saved the Whalers*

First Beach, Washington, Fall 2000

On what is now known as the Olympic Peninsula there lived an old man who was the last head whaler of the Coastal People. In his old age his wife gave birth to a son. This was their first child and the only one in the tribe who could inherit the father's whaling position.

As soon as the boy turned 12, the old man wanted to take him to sea to teach him how to whale. People from the village protested, saying that the boy was too young, but the father, knowing that he was aging, felt he could wait no longer. One morning they gathered their crew and set out to sea.

When they spotted the first whale, the crew brought the canoe up to it, and the old man, poised beside his son in the bow, sent his harpoon into the whale's back. As the rope played out, the coils grabbed the boy's leg and pulled him into the water. A long time passed as they searched frantically for him. The stricken father was blaming himself for not listening to the village people, when suddenly the boy surfaced, barely breathing, but still alive.

Plucking him out of the water, the old man and his crew rushed back to the village. There the people expressed great anger at the old man's actions. For four days and nights the boy lay unconscious. Then he arose from his bed singing a beautiful song. At the end of the song the boy told the people what happened when he was underwater: "The whale told me to crawl down the rope to him. As I did this, a bubble of air from the blow hole surrounded me. Then the whale began singing a beautiful song. I hung on and listened. When we started toward the surface, we came through different layers. Each layer was of a different color and had a different song. Each song was more beautiful than the last. Then, when we surfaced, I fell asleep."

The whale gave the boy not only his life, but also the gifts of song and color for his people. Before this time the world had neither music nor color. Guided by his experience, within four months the boy assembled his own crew and brought back the largest whale ever seen. He had become the head whaler of his tribe.

* From *Spirit of the First People: Native American Music Traditions of Washington State*. Edited by Willie Smyth and Esme Ryan, Introduction by Vi Hilbert, Seattle: University of Washington Press, 1999. Reprinted with permission.

Kwalla

The Quileute word for the mighty whale is *kwalla.* Their exquisitely woven baskets, cedar carvings, traditional button blankets, songs, and stories all portray the whale as an honored guide and ally, assuring the tribe's survival. Before the coming of European settlers, the tribe carried on a complex cultural life with its neighbors. This included trading and participating in the intricate potlatch society of Northwest coastal tribes. Potlatch refers to the ceremonial gift exchange between peoples that embodies the gracious and highly social societies along the Northwest Coast. In that potlatch spirit of exchanging gifts and cultures, we were welcomed to the Quileute Reservation on the Washington coast with a certain amount of ceremony and much generosity.

Fred Woodruff invited me to come visit his reservation, to see this living tradition of tribal canoes and to hear his tribe's vision of a new way of celebrating their ancient bond with the gray whale.

The Quileute tribe once had 10,000 members living in 27 villages, and their culture is among the oldest in the Pacific Northwest. In the mid-19th century, the white man's smallpox and diphtheria devastated the tribe until at one point the Quileutes numbered less than 800. Along with the scourge of disease, the new settlers brought a government with treaties that reduced the once-mighty tribe from 8,000,000 acres in 1855 to a reservation that is today only one spectacular square mile. They are a small tribe, the Quileutes, and they are among the poorest of all the state's tribes. And yet they have vision and courage. Restoring the tribe's language and "water ways of the ancestors," Fred Woodruff says, is their Tribal Council's hope for the future.

A tall, muscular man with a generous face and a long black braid knotted down his back, Fred Woodruff sat at the table in the tribally run cabins at La Push Ocean Beach Resort and spoke not only about the past, but also about the present and future.

In 1988, the Quileute Tribal Council passed a resolution to refrain from whaling.

"Our tribe fully supports our Makah neighbors in their treaty rights," Fred said thoughtfully. "But our Quileute elders have made a different decision. Even though we and other tribes along the coast have the same treaty rights to hunt, our elders have chosen to support the gray whale. For thousands of years, this whale has been valuable under subsistence, but now the value is in its life. The gray whale is more valuable to the Quileutes living than hunted.

"We must begin the healing here in our village and hope it can help others, as well. We Quileutes would like to offer a new vision and a different model for other tribes, as well as peoples."

Woodruff knows much about both traditional ways and new visions. Twelve years ago, he returned to the canoe journeys of his seafaring Quileute ancestors. In 1989 Woodruff and his family—wife Jill from the Yakima tribe, daughter Sharra, then 11 years old, and 10-year-old son Dakwa—joined other tribes in the first "Paddle to Seattle," a highly regarded canoe journey in which Native tribes all along the Northwest Coast, including British Columbia, participated. Then, in 1993, more than ten other Quileute tribal members completed the famous 1,200-mile round-trip canoe journey from La Push to Bella Bella, an island far north in British Columbia. Joining canoes from other Puget Sound and British Columbian tribes, Woodruff's family and other Quileute paddlers navigated along massive Vancouver Island and up to Bella Bella, where they were greeted by the Heiltsuk band of the Kwakiutl tribe.

In their family canoe, this Quileute family had paddled alongside seals and fished for huge salmon. Eagle and Raven accompanied them; they sailed on high tides and camped on beaches their ancestors had known for thousands of years.

"That Bella Bella journey had not been paddled by any other tribal canoes in over 100 years," Woodruff said. His son and daughter now share his paddle expertise. Sharra is currently 22 and her brother Dakwa, who often sits in the stern of the boat working the rudder in his father's place, is 21. "My son and daughter have been paddling tribal canoes half their lives," Woodruff said proudly. "It has given them back the ways of our ancestors with all our stories and songs and history of traveling the sea."

They have named their 33-foot cedar canoe *Os\chuck\a\bick* (pronounced Auck-la-vic) after one of the last whalers in the village who died in the early 1900s. Since 1989 Fred Woodruff and his family have journeyed over 4,000 miles in their traditional cedar canoe. Woodruff's vision for the future of his tribe and family is to evolve these tribal canoes, once used for whale hunting, into whale-watching canoes.

Quileutes have long welcomed visitors to their reservation. Karstein Boysen, the Quileute tribe's Information and Education Director, said genially, "We encourage people to visit La Push, because here is one of the few places you can just stand on the beach and see whales 50 feet offshore. The mother cows and their calves come to roll in the surf here and perhaps remove their barnacles in the sand."

Every May the tribe celebrates Quileute Elder Days, and later each summer, there is the popular Quileute Days celebration from July 15 to 18. "We have lots of singing and dancing, a First Salmon ceremony, and traditional foods," Boysen said. "It's our tribal school's way to honor the Quileute elders and surrounding tribes, too."

Quileute carver and teacher Chris Morganroth added, "We

honor the whale in our songs, drumming, and dancing. In this way, we maintain our tradition."

Fred Woodruff showed me around the reservation—from the rustic, waterfront cabins to the marina and school, much praised for its Native-language and cultural programs. Woodruff smiled as we strode down the dock toward his family canoe and the paddlers awaiting us. "We want to take people to meet the whales in our traditional canoes and along our ancestor's waterways," he said.

As if to echo Woodruff's invitation, his daughter Sharra and son Dakwa met me cordially with shy smiles and outstretched hands to help balance as I clambered into the 33-foot cedar canoe. Dakwa did the honors at the rudder. In his traditional cedar-bark hat, Dakwa could have been from another time—a time when his ancestors also paddled these arduous miles far out at sea. His sister Sharra was tall and elegant in her headdress of woven cedar. Sharra had carved and painted her long paddle herself with a bright red and black water serpent on one side and a wolf on the other.

"Whales were always welcome here," Fred explained as we settled ourselves for the paddle. "We want to share the history of the gray whale in our culture with others who care about the whale. Our ancestral coast is the only place, besides Baja birthing lagoons, where the gray whale actually lays over and rests. We have many, many whales here well into mid-June."

The other Quileute paddlers, 21-year-old Marco Black and a shy 17-year-old boy named Fawn, were the lead paddlers, their strength from many journeys much relied upon. Behind me, 26-year-old paddler Priscilla Scarborough and her 4-year-old son James were so self-assured and easy in the boat, one would hardly guess we were off on an ocean voyage. Not least among us was the Woodruff dog, Cedra, who had made many family journeys. A mixed German shepherd with a cedar-bark collar of her own, Cedra settled herself expertly in the bow of the boat in a practiced perch—her rear legs resting on the seat, her front legs on the canoe's bottom.

The sea that September day was a little rough. Woodruff had been waiting since the morning mists for the waves to "lay down a little." The dock was at the confluence of the Quileute River and the ocean—somewhat sheltered from the waves by a wall of driftwood and rocks. Behind that windbreak, the surf roared.

"Have you seen any resident gray whales recently?" I asked as we bobbed about, rearranging gear and children to find the correct ballast.

"Not yet," Dakwa answered. "Mostly migrating whales come by here. But when we do see them on our journeys, we always just let them pass. We give them our respects and just paddle alongside."

As I gazed around, it seemed that even the sea stacks, those monolithic rock formations rising out of the surf on this First Beach, looked like whales. Resembling huge dorsal fins and open, baleen mouths sculpted in stone, these craggy stacks mark the way home for this seafaring tribe. Stately in the marine mists was the sacred, cedar-clad James Island, or

A-ka-lat, where the Quileute ancestors and chiefs were buried in canoes atop everlasting trees.

The Quileute paddlers were so expert, I had the feeling that their journey was the destination. What they brought back from their travels was a living story of the highly complex tribal societies along the coast. It was a story of coexistence and sharing cultures. It wasn't about war, but about finding peace; it was about embodying the past, but also envisioning the future. Most of all, the Quileute story was about welcoming the whale.

Dakwa began a low and melodic chant, rising and falling with our paddles. The song steadied our rhythm. Echoing off the sea stacks, the break wall, the song's crescendo and fall were lyrical and haunting as Dakwa sang in his strong baritone. And we all answered, exhaling with a hearty huh-huh-huh as we heaved harder into the waves. The fatigue, the muscle aches, the fears—everything but the song fell away as Dakwa called out and we answered. The chant was in the complicated, lilting language of the Quileute ancestors, but later I found a translation:

Paddle Song

Look at the wealth of my heritage
I love my way of life

"Songs lift the spirit," Dakwa said. "We sing while we paddle long distances."

We talked about the weather and the salmon now running and returning in the coastal rivers. "Last week," Dakwa told us, "my mother caught a 48-pound salmon right at the mouth of the Quileute River over there."

Priscilla looked poised and unfazed by all our paddling work. She told us that her great grandfather, William Penn, had begun whaling at age 14 and continued traditional seal hunting from his cedar-bark canoe until 1962. That year, he set aside his sealing spear.

"Over there," Priscilla continued, without breaking rhythm in her paddle, "in the fields by the Quileute River, we used to have lots of wild horses and there are still many elk here."

The wild horses were rounded up in the late 1980s, but the elk still roam free everywhere on Quileute tribal lands. This herd is an offshoot of the famous Roosevelt elk, which Presidents Grover Cleveland and Theodore and Franklin Roosevelt each helped conserve by setting aside the Olympic National Park rain forests and wildlife. The Roosevelt elk were one of the main reasons this spectacular wilderness of Sitka spruce, western hemlock, rare ferns, cedar, and mosses have survived. Millions visit this park and the famous Hoh Rain Forest annually to see these last ancient trees. This great old-growth stands in stark contrast to the ravaged crazy-quilt of clear cuts that marks much of the surrounding Olympic Peninsula.

At last Dakwa gave the signal and we raised our paddles to rest before returning to shore. For a time, we simply floated in the sturdy cedar canoe and visited with one another.

Pink grapefruit juice was passed around and thirstily downed. It was as we drifted, paddles idle, that Dakwa told us a story: It was the wolves who first gave the Quileute people their human shape. Then the whales gave the people the beauty of brilliant colors and soul-stirring music and hunting.

As we again took up the paddles, the strong waves splashed against our sleek canoe. Again, the melodic chanting of the paddle song, and we called our *huh-huh* in return, strengthening our pace. I was mesmerized by the melody, by the blur of bright red and black designs of wolves and whales painted on our paddles.

Though my back muscles and biceps ached, I felt at one with the waves, so close to the water and diving cormorants, the floating kelp and cold splash of seawater. It was easy to fall into a reverie, along with the rhythm of the paddles: Pull back gracefully with the water cleaving in luminous wakes; lean forward and take in the mystical drape of fog over *A-ka-lat,* the ancestors' island. Our paddling was so steady and in synch, it seemed we could be journeying between worlds, the way the ancient gray whales travel between so many mysteries, both human and cetacean.

This is what it felt like, I realized, to be part of a tribal people who are still inseparable from the sea. And whenever we less-experienced paddlers faltered or had to rest, the stronger Quileute paddlers carried us, the way a dolphin or whale will carry their young, old, or weak.

At last within the shelter of the driftwood and stone wall, we felt the waves easing and we cruised back to the dock with smiles—and for some of us, with sore muscles. It was a long day on First Beach, but it was far from over.

I had been invited to a formal meeting with the Quileute Tribal Council. At the meeting Fred's brother, Russell Woodruff, the Tribal Chairman, told the story of his grandson who was learning to speak the highly complex Quileute language in grade school. "Only five or six elders still speak our language," Woodruff said. "But our children are learning it every day at all grade levels. They are also learning our traditional songs, drumming, and dances. That's how every day is started."

The Quileute language is one of only five languages in the world that have no nasal sounds (no m or n). With its complex click sounds and consonant strings, it often sounds like this: kitlayakwokwilkwolasstaxasalas. This language is distinct from any other dialect or language of the tribes surrounding the Quileutes, and has been much studied by linguists and anthropologists.

Tribal Chair Woodruff also praised his people's artists, especially the weavers, carvers, basket-makers, and dancers. To watch Quileute dancers spinning slowly in their black and red traditional button blankets embroidered with distinct tribal designs of Wolf, Raven, or Whale, their heads adorned with elegantly carved Wolf mask headdresses, is to witness "a spectacular regalia," added Chairman Woodruff.

It is a sad fact that this Quileute tribe, living among the oldest human settlements in this continent's history, is still lacking its own cultural center or museum. In fact, half of the artifacts found in the neighboring Makah museum belong to the Quileute tribe.

"We need a place to gather together and tell our stories," Chairman Woodruff said quietly, clasping his hands. Outside the modest meetinghouse, the surf off First Beach was so loud, everyone fell silent for a while. Then Woodruff added, "Our tribal paddlers have traveled these waters from northern California all the way up to Alaska. We have many stories to tell the world."

Talk at the Tribal Council meeting turned to the gray whale. Chairman Woodruff said, "We see the damage of what's taking place in Makah. Our neighbors do not own the story of the gray whales. Quileute elders have another story, another way...." Woodruff paused. Outside our modest meetinghouse, the surf swelled and rolled. "The Quileute tribe would like to declare a Welcoming the Whale spring ceremony and invite all peoples to come celebrate the gray whales on our beaches," Chairman Woodruff said. "This will bring people together in harmony, to see our tribal paddlers journey out to greet the gray whales—not with harpoons, but with songs and ceremony."

Then Fred Woodruff stood up to address his own tribe with dignity, and some measure of nervousness. It struck me how very formal they were in this official Tribal Council meeting; even between the Woodruff blood brothers. There was a sense that what they were about was nothing less than assuring their own peoples' survival.

Fred Woodruff stood up to his full height and began, "I'd like to see our community develop to become hosts to invite the whole world to see us as a model for other tribes." Woodruff paused, and then continued, "We have much to put in order as traditional people. The term 'traditional' is how we define ourselves. Interruptions have sidetracked us. We have to start getting things in order in our own village to show the world a positive picture."

Everyone nodded at the long table and the autumn ocean outside seemed to rise up, its surf wild, yet soothing. Woodruff continued, "We Quileutes are social people who bond in our gatherings, who cooperate in unity. We already have a strong vision of unity in our canoe journeys. This is a spiritual investment."

Fred Woodruff's vision for the future of his tribe includes a unique sharing of cultures. "The only way to heal ourselves," he said, "is to put back what was in place—our identity and traditions. We can call back tradition and use that as medicine to show the world, our neighbors, and all our Indian countries. My guidance comes from the echoes of my ancestors." Visitors to the Quileute will participate not only in the tribe's beautiful natural surroundings, but also in their traditions. "We will divide people into chiefs and sub-chiefs, into their own villages. They will go out and be together in the traditions, learning our songs, stories, and canoe journeying ways. Our artists will teach them carving, basket weaving, and other

Native skills. At the end of the visit, there will be a potlatch and a ceremony for all. Then we will teach the spirit of giving away to others, as it has been done for thousands of years here on the Northwest Coast."

The Quileutes hope the world will listen to their welcoming song, and take up the paddle.

The Woman Who Lived on the Island

Some people want to gaze into the future and see what it contains. I have questions about the past. Here is a story from the past. It has nearly been forgotten.

On a stormy sea, in dark turbulence, the Spanish forced all the indigenous Chumash people from an island among the Channel Islands in the Pacific. In the roundup, one woman was taken onto the ship without her child. She jumped into the water and in the powerful storm she swam the cold water, the high waves, all the way back to the island for her child.

There was a bounty for her capture, yet the island where she lived had such violent and wind-blown waters that they kept even the best fishermen from capturing her.

Eighteen years later, she was found living in a home built of gray whale bones, having created a life from the whales who had been injured, perhaps by swordfish or killer whales, and had washed in from the ocean. She was a woman as in a myth, as Jonah, who lived inside the whale. The turbulent water around her was a saving water for her.

This woman was named Juana Maria by the Spanish at one of the missions in California. Because other people were taken from the island, I think there must have been a freshwater source, also that there were other things, daily things, which helped Juana survive for all those years.

An island is a world apart. This woman on the island was a person in the cosmos, like a star. She was apart from change. The ground of her being in a place distant from change is what we all think about, trying to come to an understanding. For the 18 years she was said to have lived there without disruption, there were many changes on the mainland and her people met with those changes and were lost, as if fallen into them. California was a blood-soaked land.

Juana created a life for herself, living inside the whale. There is no mention of her child. Nor do I have her entire story. She disappeared into history as if history was, is, an ocean. The story is as surrounded by mystery as an island is by water.

Did her child live? Did she weep for years, missing her loved ones on the mainland? How many of them lived?

Her finders took her possessions. Some were placed in a museum, which was destroyed in the San Francisco earthquake and fire of 1906. A cape of sealskin and bird feathers, a cape of great beauty, was given to the Vatican.

We live inside a story. The whales are part of it. So are the people. I think of her for those 18 years watching the sunsets, the simple glow of dawn. At first I think that being tribal, she must have longed for her other people. Yet, with or without them or her child, in those years there must have been some kinds of beauty, the pleasure of watching the sea calm, the sunrise, the waves of storms, the whales blowing. There may have been dolphins about her in the deep blue water. Certainly for those taken, there were none of these simple joys. They became slaves for the Spanish, they died of epidemics, and they were beaten and tortured.

She maintained a language that was no longer known in only 18 years. In that short period of time, there was no one left alive who could speak this language. This, in itself, tells how rapidly the world had changed after contact with Europeans. And the gray whale was not only what supported her, but was one of her languages.

When she was discovered she must have known what they came for, that they came to take her to another life or death. It is said that she was found on her island sewing with a needle of whale bone. That she watched the men approach and sat still, calmly even, was a statement that she, within her being, held on to that last bit of peace she had.

Once taken from the island, she lived only seven weeks. She was buried in a mass grave with 4,000 other Chumash. She might have died of sorrow and aloneness, with no words.

Our eyes are on the past when we acquaint ourselves with a human living from the whale, a woman like the wind remembered in only a thin layer of words, but if you hear all the stories of people who became islands in the new America, her story speaks to the very heart of us in the present.

What does she say? She lived on water created, according to scientific theory, billions of years ago. Still this water is the same. At the time of her capture no gray whales were seen.

A Different People

British Columbia, Fall Migration 1999

We are sailing through the sunlit, emerald isles of British Columbia's famous Clayoquot Sound, searching for resident gray whales off the west coast of Vancouver Island. This sound is just north of Pacific Rim National Park Reserve, a national park that embraces a large marine environment, including the Queen Charlotte Islands.

"Hurry up, you," our guide, Qaamina, a Nuu-Chah-Nulth of the Ahousaht First Nations, has gently chided us to hasten away from our pristine rest on Quait Bay's floating Clayoquot Wilderness Resort and into his boat. "My brother can't breathe underwater too much longer."

Qaamina's tone is light, but his expression serious. As we circle the islands in a slow, searching glide, he explains: There are suddenly very few resident gray whales around Vancouver

Island. Qaamina teases us that his brother has to don a gray whale disguise and swim in these clear, chilled waters to imitate the disappearing gray whales.

For the past 20 years, people have come here seeking the resident grays that make up the friendliest population of whales outside of Baja birthing lagoons. Our host at the floating lodge of Clayoquot Wilderness Resort, John Caton, is also aboard. Since 1969, when Caton first met a small band of Greenpeace resisters sailing off to put their bodies between whaling industry factory boats and gray whales, Caton has followed his passion for cetaceans.

"Usually this time of year, mid–October," Caton explains, "we should see 100 whales a day just offshore in their southern migration. But now the whales are swimming 50 or 60 miles out—way out."

Jutting far into the ocean from the mainland, Vancouver Island is the first landfall the whales make in their migration to and from Mexico. These fertile and remote Clayoquot Sound waters make ideal rest stops along the gray whales' monumental migration.

Many Canadians consider the grays their "marine mascots" of the Pacific Rim; and Vancouver Island—like many points along the West Coast, from San Diego to Kodiak, Alaska—celebrates the gray whales' journey with a lively Pacific Rim Whale Festival every March and April. The Vancouver Island communities of Tofino and Ucluelet join with the nearby national park to host spring festivities, which include whale-watching, a parade complete with bagpipes, and a raft race. Year-round, Tofino, British

Columbia, continues the celebration of its nearby whale culture. The town's shops and community centers are alive with murals, artwork, and sculptures of gray whales, as the marine mammals are as warmly welcomed as visiting relatives.

Vancouver Island is also a place the grays choose to linger. Canadian biologist and author Jim Darling, Director of the West Coast Whale Research Foundation, has studied the gray whale population of 35 to 40 residents off the south-central west coast of Vancouver Island every summer since the 1970s. Darling's research shows that some adults have returned every year for the past 25 years

That is why our Native guide Qaamina's sober joke about seeing so few gray whales this past summer is meaningful. As Qaamina gracefully guides our motorboat through Grice Bay, we are bundled in our bright orange Mustang survivor suits against a sharp wind. Though it is chilly, we laze along in the golden sunlight of early autumn and keep our expectant eyes on the waves.

"I've seen only one juvenile gray here in the past several weeks," John said.

Qaamina and John both fell silent. There is an air of contemplative sadness about the two men as we steer slowly through shimmering kelp forests and eel grass. "Maybe soon," Qaamina said softly.

At 38, our Nuu-Chah-Nulth guide seems as familiar and concerned with gray whales as with his own people.

Clayoquot Sound was named for the band of Nuu-Chah-Nulth who first lived here. According to one source, Clayoquot means, "another or a different people." This definition seems to suggest a people apart or different from the European immigrants. But another translation of the word is more haunting: "people who are different from who they used to be."

As we motored slowly past Meares and then Flores Islands, home of Qaamina's Ahousaht nation, Qaamina said softly, "No whales...yet."

Canada's Indian nations, now more properly termed First Nations, have a long and proud history. Qaamina's Ahousaht band of Nuu-Chah-Nulth are believed to be the elder peoples of this far Northwest territory. Opitsat, the principal village on Meares Island, has archeological evidence of occupation dating back 6,000 to 10,000 years. Such findings support the claim that this is the oldest continuously inhabited settlement in North America. Such authority of tradition and territorial longevity lends the Nuu-Chah-Nulth people a sense of power and perspective.

Qaamina's father, Stanley M. Sam, Sr., of the Ahousaht First Nations, was named Momoquin by his grandfather, a man whose teachings took the form of moral storytelling with animals such as Raven, Wolf, Deer, Salmon, and Whale as main characters. These stories are now passed down to Qaamina, as well as to the world in several books. Stanley M. Sam, Sr., has written eloquently about the whaling tradition of his ancestors—a tradition that in the past was very much in harmony and in a balanced relationship to the gray whale.

"In all my years—20 of them—watching the grays," Qaamina said at last, as we motored through flat, quiet waters,

"I've never seen a summer this bad for the whales."

In his black cowboy hat and graying beard, John scanned the waters. "Just this past spring, we had the most whales I've ever seen in all my time here. At least 59 or 60 resident whales. Sometimes we'd just float out in the middle of two dozen mother-calf pairs in a little nursery all our own. The Friendlies would come to our boats and we could even touch them. Whales everywhere!" Caton hesitated a long time, and then added, "I wonder now if we'll ever see that kind of trust again."

"We've seen some amazing things out here with the whales," John took up again, almost having to shout above the engine as Qaamina decided to take us farther into a labyrinth of little islands and inlets. "Last spring in April, I swear there were two dozen mothers and calves right near the resort. We put on dry suits and dove with them, trying to keep our respectful distance." John grinned and said, "That is the most thrilling thing that ever happened to me in my life...floating among them in the kelp forest, gray whales everywhere. Then suddenly out of nowhere a huge gray surfaced near me, so close I could put my hand on its nose. We were eyeball-to-eyeball!"

Hearing this, Qaamina gave out another of his rare smiles and scanned the horizon more hopefully—as if the story might summon the gray whales home. "We took a lot of people whale-watching," Qaamina said with a nod. "Lot of grays and a lot of happy people."

"Other amazing things I've seen about the grays," John

again took up. "Off Wilf Rock, there was a gray whale mother and her baby just floating easily. Suddenly there were orca fins all around them. But the mother rolled over and actually *lifted* her calf onto her belly. This meant that she couldn't breathe for long because her blowhole was now submerged." John stopped, marveling. "She was sacrificing herself for her calf. And you know what? Those orcas up and left."

John's story was soon followed by a shout from Qaamina, whose steady eyes were fixed ahead: There we sighted a very small, juvenile gray we estimated at one to two years old. As always, my first thought was that perhaps this young gray was White Beak, for whom we had been searching ever since we encountered him at his birth in Baja's San Ignacio Lagoon. But where White Beak was almost albino along his snout, this gray was dark and mottled. We could see through the binoculars that barnacles and sea lice were just beginning to adorn his dorsal hump and mouth. As he slid gracefully into the waves, I tried to memorize his distinctive knobby patterns of "knuckles" along his dorsal ridge.

Linda noticed that this whale had a slight notch in his tail flukes so we called him "Tail-Notch." Qaamina kept us a respectful distance from the juvenile whale and cut his motors to let the current glide us alongside. The young gray spyhopped curiously and took us in with that big eye.

What must we look like to Tail Notch, I wondered? In our orange fluorescent survival suits, hoods, and sun goggles we must seem like astronauts from another planet. The gray whale lingered with us about a half hour, accepting our presence, as he dove and fed on the rich nutrient bottom of the Sound.

We all speculated about whether this was a resident gray. "Hope so," John said. "Maybe he'll stick around."

We stayed with the young gray a little longer, until he at last dove with a gracious tail fluke up as if to say goodbye. After the gray whale's departure, I was suddenly aware of how chilly it was out on the water, even though this October sunlight was still bright and beckoning.

At the end of our long day on Clayoquot Sound, having encountered only one gray whale, Qaamina turned the boat around for the long ride back to the wilderness resort. With its lovely isolation, snuggled in Quait Bay against an old-growth cedar rain forest, the Clayoquot Wilderness Resort is one of my favorite retreats. This spacious, two-story cedar-wood houseboat is accessible only by floatplane, boat, or helicopter.

Once back at the floating resort, we climbed out of our bulky, bright orange survival suits and settled next to a roaring October fireplace for some hot chocolate and conversation. Without the noise of the outboard engine or the distraction of the pristine scenery and possibility of seeing gray whales, it was as if we were meeting each other for the first time.

Qaamina was relaxed and surprised us, after his reserve on the boat, by engaging in a dialogue that ranged from his father, the Ahousaht Nation historian, to his own children. It was as if he had been waiting all day to truly talk to us.

"I want to pass down to my own children," he began

intently, "what I have learned from my father and he learned from his father—stories that teach us everything we need to know about where we are from and who we are with here. All the animals..." he nodded firmly. Then he continued, leaning forward "We treat animals the same way we want to be treated," Qaamina said in his rhythmic, deep voice. "Our cedar forests and gray whales are sacred."

Qaamina's family was one of the first to begin whale-watching in Clayoquot Sound. Whale-watching off Vancouver Island has been a source of spiritual relationship and family pride for the Ahousahts and several other First Nations people who have lived here for so many thousands of years.

It was Qaamina who brought up the Makah whale hunt. "It was 6:49 a.m. on that morning last May," Qaamina said, his face troubled. "I turned on the television like everybody else, and there it was...the gray whale coming to greet the people in their canoe...the way our Friendlies here do," Qaamina paused and looked down, shaking his head. "Then they struck that young whale with a harpoon, then those big rifles. And I had a very dark feeling."

Qaamina did what he always does when he wants guidance—he called his Elder father. "My father told me, 'This was a day of reckoning that had to come. This was not a spiritual hunt; this was not what it used to be.'"

His father explained to Qaamina that in the old whaling days of his Nuu-Chah-Nulth ancestors, no one ate the gray whale. "'That is not the right whale to kill,' my father said. 'Traditionally, we ate the humpback.'" Qaamina fell silent, then continued with a frown on his usually genial face. "My father also told me this: 'To be a great hunter, you don't hunt the timid whale.'"

We all were quiet as if in memoriam of the whale taken that morning in May. Qaamina added, "I felt so bad after that whale hunt I had to put it into drawings that I shared with my Dad and family. My cousin said, 'Can I take these for a while?' And he just went down and faxed all my drawings to the Makah Tribal Council."

Qaamina sighed, "You know, I have relatives in Neah Bay and I studied engraving with my uncle there who taught me so much. But I do *not* agree with this whale hunt . . . they sacrificed the whale to show the world they had rights. When they called me from Neah Bay after some of my drawings and words got put in the papers, they asked me, 'Why do you mock us?' And I told them, 'You mock yourself.'"

Qaamina was thoughtful and rested his head on his hands, which are at once the rugged hands of a lifelong fisherman and also tapered and elegant, the hands of an artist and silversmith. "That whale hunt has put distance between me and my uncle in Neah Bay. It put a border between us that was never there before. There was never any border between the Indian peoples of the U.S. and us until now."

As if to underscore that division between Canadian First Nations and U.S. tribes, the First Nationals Environmental Network in Canada criticized the Makah hunt by running a protest on the Internet:

> Not all indigenous people support Makah whaling. We

> are deeply concerned and saddened by the killing of a whale at Neah Bay, Washington by members of the Makah Nation....Japan, Norway, Iceland and other countries are working towards getting commercial whaling approved once again. The Japanese have been lobbying First Nations Peoples on the West Coast and around the world to open the door on "cultural whaling," which they also claim as a "right." The Makah Nation is divided within, with many elders and others speaking against this "return to traditional practices" and their voices are being ignored and suppressed. While we respect Treaty Rights, this is political reason being used for killing and not a true meaning of need when it comes to the taking of another being's life. Using Treaty Rights in this way may set dangerous precedents.

Qaamina also spoke to the Canadian newspapers, decrying the Makah hunt. To this day he is surprised that few, if any, U.S. reporters called for reactions from other First Nations people outside the United States. "Maybe your press down there don't want to know that some of us here are against it," he surmised. "But everyone was talking about it up here among the Nuu-Chah-Nulth," he said.

Qaamina believes that perhaps the peoples of the whale were also communicating about the hunt.

"At 6:49 a.m. a message went out from that gray whale to all the others," he said thoughtfully. "My father, the traditional Elder who was on the science council, told me that whales are intelligent and communicate over long distances. When they grieve it's like us grieving lost people we love." Qaamina fell quiet, then at last concluded, "When that whale was killed it was probably already giving out a message to the other whales."

"Yeah," John echoed. "I've been around whales enough to know that all mammals communicate, too. Look in their eyes—they are sensitive, sentient. And when that whale gave out a message it was about us as humans—*Don't trust us!*"

"Today we saw one whale in Ahousaht Bay," Qaamina all but whispered and sat back, putting his coffee cup down on the table with a sense of resignation. "Usually we see at least nine...."

"And sometimes as many as 25," Johan added.

"So now we have to tell people who come—not so many whales maybe," Qaamina said.

"I never thought this would be possible," John said.

Qaamina stood to gravely shake our hands and ask us to be sure to tell his story in our book. "Things are going to happen now, that's what my father says. He also says, 'This was not a great hunt. A great hunter will not go out and shoot his dog and take it home for dinner. We wouldn't kill our whales.'" Right before he took his leave, Qaamina jammed his artist's hands in his jacket and admitted, "I still want to be close with my Makah relatives, but I have to let them know what I think. We have generations to come behind us. I want to make sure my children never harm a gray whale, not for the next four generations."

The Whaler's Shrine

When the white men arrived on Vancouver Island, some of them described the carvings and the life-ways of the tribes. An article by Aldona Jonaitis describes the Mowachaht Whaler's Shrine that once existed along the west edge of Vancouver Island, in Nootka Sound. While the people there may not have hunted the gray whales, the grays passed by this sacred place. Some European recorders of history say this was a place where slaves were sacrificed. Some say it was a place to worship ancestors or to pray for whaling fortune, good luck in the hunt.

The whales passed by this sacred, well-hidden place for many generations before the shrine was dismantled by anthropologists and taken to a museum in New York. A photograph of the shrine as it stood here reveals hauntingly beautiful carvings of wooden people standing in rows around the enclosure. They survived the winds and rains of time. They are luminous and powerful in the photographs.

One looks as if it is singing. Jonaitis thinks some of the figures were created to fit into spirit canoes, to replicate the process of killing a whale. Young Doctor, the Makah carver, made small figures with spirit canoes. Perhaps this is why she came to this conclusion. I think of the wooden beings covered at times with frost, perhaps algae, yet they survived in that place with strong winds, in the smell of cedar and fir, in a place of many mountains, rivers, and lakes.

The people in the photograph are beautiful and haunting. "Why do they stare eerily out through eyes with no eyeballs, penetrating the viewers with a depthless gaze?" Joanaitis writes. Four human skulls are lined up along the ground and in the center are two carved whales. Whale bones were also said to be in the shrine.

The indigenous view of this sacred place was finally taken into account recently when the Mowachahts wrote their own statement, calling the shrine the Whaler's Washing House. As with all tribes, the numbers of the Mowachaht, a band of the Nuu-Chah-Nulth people, severely declined with white contact. In the first decades of European contact, a population of several thousand people was reduced to less than 300 and since then the population has remained about the same.

The First Nations people at Yuquot, in British Columbia, are now returning the ancestral shrine to its home. In the past this Yuquot village was the capital for 17 tribes called the Nootka people. The area was a primary center of trade, as was Cape Flattery where the Makah are still located.

The word "Nootka," incidentally, was not a name for the people as some think. It was the word used to direct Captain Cook to a safe place in order to anchor his craft. It meant "to go around the rock." This name, as in other place names, was given to the people.

Indian whaling, as one Nuu-Chah-Nulth man recently told us, was once a sacred and dangerous calling. Their canoes had to come from a perfect, well-chosen tree. As it was being made, the makers sang to the tree. Whaling was a precarious occupation and the whaleboat needed to be precise. It had to float in perfect balance or it wouldn't hold the hunters safe. Perfect balance, like the minds and spirits and hearts of the whaling men were supposed to be. The paddles were made of yew or maple, tapered perfectly to lessen the sound and greaten the speed of them as they entered the water. The harpoons in this region were 18 feet long, 4 inches in diameter, and made of yew wood. The tip was a mussel shell and elk horn, the sinew covered with bark. The entire thing had to remain rigid. With the rest of the gear was the lance, which would inflict the last wound, killing the whale as humanely as possible.

A New Century of Whaling

We move now into a 21st century of global whaling practices and possibilities. Many conservationists dread the probable return to commercial whale hunting worldwide if pro-whaling nations succeed in delisting gray whales and again begin hunting them. Japan, as mentioned, has already expanded its scientific whale hunting to include rare and endangered species, such as the Bryde's and sperm whales. Environmentalists and others believe that it is critical to convince Japan to adopt a policy of environmental stewardship of the gray whales and to honor the international ban on hunting gray whales.

A major part of whaling in the 21st century may well involve indigenous peoples returning to whaling. The issue of resident gray whales is vitally important because those populations are small and unstable. Internationally the power struggle continues between pro-whaling nations who are determined to convince the IWC to vote for a future of commercial whaling and those countries working to protect whales world-wide. Indigineous peoples may well be the key players in the politics of whaling. Or they may again be the victims.

Canadian biologist Jim Darling adds a cautionary note as governments try to manage the gray whale populations of the future. He explains: "Once there were four populations or herds of gray whales in the world. Then there were three, now there are two—just barely. Only one is healthy; the other is hanging on precariously."

In this 21st century, if indigenous peoples again return to their whale hunt traditions, supported by the very Japanese and Europeans who in the 1800s proved themselves so rapacious, what will be the fate of the great whales?

Perhaps we need a new way to look at other species, now that our own subsistence is not so directly linked to hunting. Who is speaking today for the rights of whales? And should their rights be considered as deeply as the cultural rights of any people?

Commentator Charles P. Fall of public radio station KBOO in Portland, Oregon, asks this question: "Do the Makah speak for nature or for their culture when they exercise their treaty rights to hunt the whale?... What we are contending with today is the effect human 'second nature' is having on nonhuman 'first nature.' ...Not even the Makah are immune from rendering an ecological impact statement of their deeds, of what they do.... This leads me to ask, what vision could unite Native Americans, whites, Asians, Africans, and all creatures on earth into a meaningful unity?"

PART FIVE

A Cold Eye: The Far North

Alaska, 2000-2002

Sea of Mystery

The white men called the Bering Sea "the Sea of Mystery." This rich green sea with eel grass beds seemed ghostly and eerie to those unfamiliar with the place. It held shoals, fog, a clouded world shrouded as if secret acts could take place there, acts hidden even from the eyes of the creator, of all creation. In summer the shallow waters were filled with life. In winter it was frozen, with islands of ice drifting off and pack ice filling in. The Bering coast came to be called "Whale Alley." Whaling countries fought for their hold in this place that belonged to none of them. In 1867 the United States bought Alaska from the Russians and took possession of whaling rights without considering the Native peoples.

Because whales were valuable for their oil, the Bering and Chukchi Seas were soon filled with ships and the whale populations decreased dramatically. Great lives, once breaching, once breathing, *were gone. They were harpooned near the Aleutian Islands and Unimak Island. One whaling ship, the* Kamchatka, *was so soaked with whale oil that it caught fire. Canoes with survivors of the* Kamchatka *reached islands near the Aleutians.*

In the Russian arctic, Eskimos who worshipped and hunted the whale were forced to become butchers for the Russian factory ships in the 1920s. No ceremonies were allowed and the Eskimo shamans were put in jail. The people were reduced to cutting up gray whale meat to feed the caged foxes Russians raised for fur.

Now, in a 21st century with Russian people also starving, the Eskimos in the Russian Arctic depend again on subsistence whaling. But the hunters are telling scientists that the gray whales are too skinny these years. Their meat is so rancid it smells like medicine. They won't even feed it to sled dogs.

Scientists come to the far north to talk with the old people who have always lived on the ice. They hear stories of walrus and polar bears wandering too far away to hunt because the ice is shrinking. Toxicologists come and take samples of Eskimo women's breast milk to measure levels of PCBs, contaminants from the toxic blubber of marine mammals. But the scientists don't tell the Eskimos what they've found. They only say that sometime soon, the Bering Sea will be open ocean where there used to be ice.

An Eskimo, Caleb Pungowiyi, tells scientists who come to talk with the elders about their knowledge of this icy world, "When this Earth starts to be destroyed, we feel it."

Blue Ice Mountain

Old Northwestern Glacier, Alaska, Summer 2000

No grays seen as we sail all day through the ice-clotted Kenai Fjords, but other great beings float, breech, and spy-hop on the frigid waters among tidewater glaciers named Sunlight, Southwestern, Anchor, and our destination, Old Northwestern. Perhaps it was all the rugged ice floes scraping our bow, but everyone shivering on board the small motorboat was suddenly subdued.

It was as if we had reached the end of the world and the beginning of another planet made of massive ice majestically poised and buoyed by salt water. Gazing up at the azure ice of Old Northwestern glacier, I felt complete calm.

Water and ice shape this world. And it is here in the frigid, fertile waters of the far north that the gray whales find their food, their summer sustenance—the delicate millions of amphipods and other crustaceans. In all our travels, we had never seen the grays foraging, their huge mouths shoveling ocean bottoms, their baleen straining through mud to filter the tiny, teeming lives that feed such a large creature. We had also never once seen any whale we could truly identify from our seasons in Baja lagoons. This was not a first trip to Alaska for Linda or myself, but it was the most memorable. The grays we discovered here in Arctic waters made it so.

Deep in the Kenai Fjords, the world of summer grass and bright sun seemed far away. Here was the kingdom of frosty mists and seemingly endless ice. It is strangely restful to be in the presence of natural power so far beyond anything measured by humans. Ancient rock, mammoth ice floes frozen and floating like continents adrift in time, these tidewater glaciers are tinged an otherworldly turquoise because their density does not absorb blue light. So as we eased our way through waters afloat with ice slabs, Old Northwestern Glacier loomed true north like a pale, blue ice mountain.

No one could say a word. Our skipper cut the engine and we rocked like a small bobbing cork. We were quiet, but Old Northwestern was not. Ice calves clinked and a waterfall thrummed down nearby green slopes, which as recently as the 1940s were still part of this glacier. The bottom ledge of Old Northwestern looked like a million years of blizzards thrust up by a gigantic snowplow into a bright bank afloat on the salt water.

Then it happened, what we had all hoped for and held our breath to hear: The ancient glacier calved with the mighty sound of worlds colliding, the earth giving birth. Groaning, the glacier cleaved off an iceberg—literally a huge ice-chip off the old hoarfrost block. One had only to witness this calving glacier to remember that 15,000 years ago, an ice age gripped our planet.

Gone now was summer's balm as we rocked in silence, except for the clank and crack of ice floes against our fiberglass boat. In Old Northwestern's chilly blue shadow, the temperature had dropped a good 20 degrees to freezing.

"As mighty as this glacier is," our naturalist said in almost a whisper, "it's still sobering to remember that since 1900, Old Northwestern melted away by ten miles."

Some scientists say that the Arctic sea ice is decreasing at an alarming rate. Since 1978 it has retreated by 6 percent, an area roughly the size of Texas. A 2002 report on global warming released by Environmental News Network (ENN) notes that the U.S. Geological Survey (USGS) measures the world's seas as rising "nearly one-tenth of an inch each year, fed by rivers of melting glaciers and ice sheets around the globe." If the seas continue to rise at this rate, the USGS predicts, Iceland's glaciers could disappear by 2200 and "that event alone could raise sea levels by 20 feet." The ENN special report also warns that the massive ice fields topping Africa's Mount Kilimanjaro, first mapped in 1912, have lost 82 percent of their mass. In Peru, the Quelccaya ice has shrunk 20 percent since 1963. These signs of global warming, scientists say, show climate changes in fragile, high glaciers that will affect everything from our environment to our economies. Changes in glaciers, they say, are among the first indicators of a natural world gone awry—and probably, the scientists conclude, much of the damage is man-made. But when the ice changes, so must we.

Some scientists warn that Arctic ice is melting so fast it could disappear entirely each summer beginning in just 50 years. This prediction is supported by oral histories and observations of Siberian Yupik hunters in Chukotka, who in the spring of 2002 have had to expand their many words for ice and snow to include *misullijuq* (rainy snow) instead of simply *umughagek* (ice that is safe to walk on).

Because sea ice is now covering 15 percent less of the Arctic Ocean than it did 20 years ago, the vast ice masses upon which walruses, polar bears, and seals nurse their young are unable to sustain the Arctic animals and hunters who hunt them. In April 2002, Inuit people in Nunavut, Canada told interviewers that the weather was "*uggianaqtuq*—like a family friend acting strangely."

This change in the Arctic world profoundly affects the gray whales who summer here amidst calving glaciers and unpredictable ice. Shallow, muddy ocean bottoms of the Bering Strait are the gray whales' most important summer feeding grounds. Amphipods—those tiny, red-and-white-striped shrimp-like creatures—provide most of the energy for the grays' long migrations. Some scientists believe that the warming currents of the late 1990s El Niño cut into the Bering Strait's food chain. Recent satellite data show a shift in plankton species; those that usually sank to nourish amphipods instead floated at the surface of the Bering Sea. But if this crash in the slow-growing amphipod populations was responsible for the gray whale die-off, then why has there been a slight rebound in whale births and fewer die-offs in 2001 and 2002 migrations?

A new theory brought forth in the spring of 2002 suggests that perhaps it is Arctic ice that made the difference between a healthy migration and a record die-off among grays. Scientists note that in 1998 and 1999, the spring Arctic ice broke up quite late in the Bering Strait. This might have severely shortened the grays' feeding time and forced them on their migration weakened and malnourished. Hundreds of gray whales may have also died, unseen, out at sea. But in the 2000 and 2001 seasons, the Bering Strait ice retreated much earlier, allowing the grays to feed their fill—thus, perhaps assuring a healthier migration and fewer deaths. Is Arctic ice really the answer to the great, gray whale die-off? It's still a mystery. As with any unexplained event, we must turn not only to scientists' theories, but also to our own stories and imaginations in contemplating a changing Arctic world.

Norwegian polar researcher Tore Furevik writes that ice-dwelling mammals, and therefore indigenous peoples of the far north, may be making a desperate last stand in the not-so-distant future. "When our land and animals are poisoned, so are we," says Inuit activist Sheila Watt-Cloutier. "We are the land and the land is us."

In the shadow of Old Northwestern Glacier, it was hard to imagine that this pristine last frontier was also so deeply disturbed. It was even harder to fathom this world without Arctic ice.

None of us knew that only two months later, in late August 2000, a Russian icebreaker would visit the North Pole and discover open water instead of ice. Some of the Arctic experts are taking a cautious view of its ultimate significance—most of the ice-measuring stations in the Arctic have records dating back only 50 years. That is why it is so important for scientists to listen to what the Native peoples are telling us about the change in ice and animal behavior. It used to be that when Eskimo elders in Yanrakynnot of the Russian Arctic told stories of the ice melting before summer, the village children would laugh. How could this ever happen? Their homeland was frozen almost year-round and the ice securing the Bering Strait was so thick, hunters could guide sleds heavy with whale carcasses.

Now as the Arctic ice is getting thinner and the tundra is melting, reindeer herds are dying and strange shrubs and mushrooms are blooming. All these are unsettling signs of global warming. While we do not know the long-term effects of global warming, we do know that Arctic temperatures at the end of the 20th century were the warmest in the last four centuries. A recent World Wildlife Fund (WWF) estimate theorizes that 20 percent of all living species might die out due to shrinking habitat in the Arctic and northern latitudes. Even a change of one of two degrees can change migration patterns and the ability of some plant or animal species to adapt. Shrinking ice and warming waters, says the WWF report, may well throw such species into a "race for their lives, and scientists do not yet know whether they would be able to migrate fast enough to outrun the change." The WWF report estimates that some species of vegetation might have to migrate ten times faster than during the last glacial retreat.

I faced Old Northwestern's mammoth, frozen flow. Aided

by such natural events as El Niño cycles, are we creating a dangerous warming trend with our fossil fuels and greenhouse gases? The U.S. is the largest single source of fossil-fuel-related CO_2 emissions, with an all-time high in 1997. In 2002, the Bush administration was still backing away from any commitment to sign the Kyoto Treaty, which would limit our greenhouse gases. We wait for politicians and scientists to figure it out and end their arguments—even as the Arctic ice melts and the world grows warmer.

As our small boat bobbed about in the iceberg-laden bay beside a towering, turquoise Old Northwestern Glacier, I contemplated this ice that is so remote, yet vital for the world's health. It was hard to comprehend that so massive a glacier as this could yet be in retreat.

I stood on the deck, trembling with the cold as Old Northwestern Glacier breathed a frigid pall over our boat. In this intimidating, icy realm I tried to imagine what might counterbalance our human willfulness and wanton disregard for nature's way. Watching the ice floes crack against the sides of our boat, I remembered a story that had much to do with the beginnings of this book and our journey following the gray whales.

Ice, Ice Hole, and Snowflake

Point Barrow, Alaska, 1988

Inupiat hunters in remote Point Barrow first put gray whales on the map of the world's imagination in October 1988. As often happens in the far north when encroaching ice floes seize the Arctic waters, three gray whales became trapped in small breathing holes, where they would likely languish until they drowned. Four miles from open water, Inupiat whale hunter Roy Ahmaogak spotted the trio of trapped whales and reported their plight to a local news station.

Poignant images of three bruised snouts, bloodied as the whales struggled to surface and breathe every four minutes in their shrinking water hole, mesmerized an international audience. From all over the world, media gathered on the ice to report daily on the progress of chainsaw-wielding Eskimos, tireless Greenpeace coordinator Cindy Lawrence, and U.S. military personnel all engaged in an unprecedented and massive effort to save these three whales.

Once the whales were named—Bonnet, Crossbeak, and Bone—there was no turning away from their plight. The Eskimos, who struggled to chop more breathing holes with their chain saws, also named the whales—Putu (Ice Hole), Siku (Ice), and Kanik (Snowflake). The Inupiat, subsistence hunters of the bowhead, have never hunted the gray whale. As the whales' breathing holes began to slush over and the rescue effort seemed doomed, unexpected allies joined the drama. Two brothers-in-law from Lake St. Croix, Minnesota, Rick Skulzacek and Greg Ferrian, had watched the desperate whales on the evening news. Immediately they loaded up their family firm's electric de-icers, flew to Point Barrow, and set about churning the warmer water up from below to stop the ice from closing off the whale's only source of air.

The survival saga of Putu, Siku, and Kanik was daily news. A world watched as Kanik, the youngest gray, struggled to breathe. The other two, Putu and Siku, seemed stronger.

Another unlikely hero was author and musician Jim Nollman, who has spent decades making music to interact and communicate with cetaceans around the world. As Nollman played a small white electric guitar through hydrophones dropped into the icy water, he hoped to attract the three whales away from their familiar breathing hole and toward the other breathing holes the Eskimos carved nearer a channel of open water. The key of D major has often been a favorite among cetaceans, Nollman noted, but it didn't seem to stir the whales from the breathing hole that would close up overnight, drowning them. In desperation, he played a tape of the South African group Ladysmith Black Mambazo, the soft, harmonic humming of a people a world away from these whales. Suddenly, the whales dove. And when they resurfaced, they had left the shrinking breathing hole to find others nearer open water. Everyone shouted in disbelief and joy as the whales continued between carved ice holes and nearer the channel a Russian icebreaker was opening to the sea.

As everyone cheered, it took a moment to notice that the third whale, the one called Bone or Kanik, was not with the others. Perhaps the liberating swim from ice holes to channel was just too much for her weary, young body. She had drowned.

But Putu and Siku were free and over the next three days they followed the Soviet icebreaker to open sea. At that moment of interspecies celebration, no one reported on the fact that Russia was then and still remains the world's largest killer of gray whales. No one talked about the fact that this drama of ice-choked whales is a story that happens off-stage almost every fall. All of this was put aside—because for this one, brief and beautiful moment, humans and whales worked together to show the trust and altruism that is possible between our species.

The Floating World

The life of a gray whale consists of the small and minute. If you could see diatoms with the naked eye, they would look like houses of crystal, boxes with exquisite patterns. Their formation is one of the mysteries of the Earth. Beautiful, shapely, at times some of them form a film over the bodies of gray whales and make the skin of the whale shine luminescent in the dark so that its passages north are covered in beauty in a floating world both delicate and powerful.

They Follow Light

Light and water become organic matter. In large numbers, unseen by the naked eye, are living cells that the whales follow in their journeys. During our winters, the water in the desert places such as San Ignacio, Scammon's Lagoon, and Magdalena Bay is shallow and warmer than other parts of the Pacific. Because there is sunlight, the water is rich, even thick, silted full with plankton and crustaceans.

Even richer are the summer seas in the north where sunlight is a constant. The males leave the southern regions in Baja first around February and travel north 100 miles a day, and in this journey, they know by feel the swells and currents, the face of light, and the forces of gravity pulling them northward. It is an elemental geography, that of the gray whale, with its underwater dunes and reefs, with the magnetite in their brain pulling them toward the north.

The females and young follow the males north. In the wake of water, even barnacle-encrusted and back-knuckled, they are shining and dark when they breach.

Plankton

The word "plankton" derives its meaning from the word for floating or wandering. This is how plankton moves in the ocean world. It drifts. It rises and falls in a vertical migration. The plankton in the water looks like a snow, softly falling, each one different from the others, like snowflakes. Each ounce of ocean water holds immeasurable thousands of lives.

Among the plankton forms are plant cells. The microscopic plants bloom in bewildering profusion. They are forests of the ocean, moving across the ocean sometimes like a storm, especially where temperatures change, and sometimes settling to the bottom where the whales feed.

It is a system of exchanges. The gray whales dredge up the sea bottom, creating richer silt, a more sunlit plankton. When exposed to sunlight the plants bloom. Through photosynthesis, they not only *support the life of the ocean but provide 80 percent of the Earth's oxygen. This means that our lives are enmeshed with those of the whales in ways we don't usually consider. Our air, our breath, begins there, drifting through an ocean, in its rich and richly inhabited suspension of life.*

The plankton drifts, blooming, with the sun. In the coastal margins of the sea, clouds of plankton drift into accumulations in bays and inlets and feed tiny crustaceans, shrimp, krill, and zooplankton—single-celled animals, minute carnivores, which themselves feed upon the plant plankton.

The plankton are plentiful in places the whales know by feel, places where waters meet with other waters, where the temperature changes, such as the water over volcanic vents, and where freshwater springs enter a larger sea. Whales find their ways into bays and inlets where the plankton has settled and they take into their mouths the rich bottom of the sea, sometimes over a hundred pounds a mouthful. They then force the water out through their plates of baleen, trapping plankton, small animals such as krill, phytoplankton, copepods, tiny crustaceans, and small shrimp. As the whales disturb the bottom, plankton are again silted upward to the surface. It is a relationship—the whale, the sunlight, and the plankton—a triad in which many transformations take place.

The whales pass the land once belonging to large numbers of Chumash peoples. Following the "coastal trough," the trench along the West Coast from Baja north, they pass Trinidad, California, where the Earth is saturated with whale fat from the days when it was a large whaling port. They move past the volcano erupting now off the Oregon coast, feeling its temperature changes, its warmth on their barnacled flesh. Following the landscape along the coast, they see tree stumps beneath the ocean, dated from the time of Christ. They are stunning, these stumps, revealed last year on the Oregon coast. This was a forest unknown, never remembered, until an earthquake changed the layout of the coast.

They go past La Push, Washington, where the Quileute live, with their beautiful cliffs and the time-softened beach of circular stones washed by time and the sea. On land, wind moves the sand.

The whales pass the old whaling village of Ozette and Cape Flattery, another of the large whaling posts. They know the upwelling of spring waters, the places where rivers enter the ocean.

They pass the Whalers' Shrine where once the traditional peoples of Washington went to pray and cleanse themselves for whale hunting. And everywhere they dive to the bottom to scoop up sediment. They create a spinning of plankton; it is drawn to the surface. It is changed by the sun.

Along the coastal waters where we have followed the whale migrations, we know only the surface world. We see the surface. The whales have a different sighting, they have a life beneath, a vision at the depths.

In these depths are organisms even smaller, though not lesser, than the plankton. The drifting plants, the phytoplankton, and the zooplankton are all made up of diatoms. Diatoms store food in small droplets of fat. By and by, consumed, these become the fat of the whale. The whale is fattened by these motes of life.

In their passages, in their world beneath, the whales also travel near and through kelp beds. These are forests of the ocean, anchored to the ocean floor, bending, the tall plants moving with currents and tides. In them, the whales protect their young. They sleep in these forests, floating at the surface, rising every so often to breathe, and even have been known to snore.

Once while kayaking in the beautiful kelp beds at Friday Harbor, watching killer whales, the gray's enemy, surfacing and blowing all around us, I looked down from above. In the green light I saw small white jellyfish the size of a child's fingernail. And once I saw an octopus, pale there in the green suspension of light and water shared with whales.

Following their brain compass, following the light, the grays pass through the strongest currents on the West Coast, in water cold and marked by riptides. They pass through storms and near black rock beaches. Their world is one of constant movement and change. Snow mountains are in the

background. Even in summer there are cold winds, the sharp cold rise of land.

The whales pass by the green of Kodiak Island, where bald eagles are abundant on the trees, the Chugach Mountains are in the distance, green, and dark, snow-covered ranges rise in the background.

We know where they are by our maps, our own named places, the Aleutian Islands; Ugak Pass; St. Lawrence Island where the whales are treated with reverence; Chatham Strait. These are places the whales know by the compass in the brain, which pulls them north.

And also, unheard by us, they follow sound. In addition to what the whales feel and see, the pull of sunlight and gravity, there are languages in the ocean: the clicking of coral, the voices of other whales, other species, including plants. They hear the tides rising, the great energy of gales of wind. Not only is it a place liquid, it is a place alive and wild with sound. Sounding through the water, hearing the echo returned, this is part of their mystery. Whale bones carry sound as if their body, itself, is an ear.

As the gray whales approach the northern seas, the ice melts and disappears before them. In their return journey south the ice will form behind them, as if to chase them away. But now, after traveling around the Aleutian Islands, to summer in the Bering and Chukchi Seas, they feed in waters green and rich with plankton charged by the 24-hour northern light. There, the waters are green with chlorophyll and the sea bottom is rich. The whale eats.

We've traveled with the grays, Brenda and I, wherever they've gone, following them in kayaks, small planes, boats, and ferries. Flying in a four-seater, we go to Tofino, British Columbia. Everything this year is different, according to the First Nations people we've talked with, perhaps because of the whale hunt, as some Alaska and Canada Natives think, or because of the changing sea; for whatever reason, their migration has changed. The whales are usually found in bays, inlets, and coves at places along the coast named Clayoquot Sound, Ucluelet, Esowista Peninsula, Opitsaht, Ahousaht, places with Native names; but in these places now the whales have disappeared. Qaamina says they disappeared on May 17, the day the first whale was killed further south. It could be that their trust for humans no longer exists, he tells us.

The whales have remained farther out from the shore, unable to eat the food they need, food touched by the sun. They are seen far out by tuna fishers, and the whales are traveling faster than usual and losing weight.

In Kodiak, near cliffs with fossils, in a place of nearly all-day sunlight we meet up with Erik Stirrup and Stacy Studebaker. At teacher Stacy Studebaker's table, I look through the microscope at a single drop of water and see life-forms moving. There are 500,000 organisms per gallon of Alaska water. Zooplankton consume phytoplankton before my eyes, and I yet can see the swallowed within; they are see-through, shining, like crystals, like amber.

The next day Erik takes us out in his boat, 7 1/2 miles beyond Ugak Pass. We are in the midst of gray whales. Suddenly in the background I see a blow unlike the heart-shaped breath of the gray. This one is a column going straight up into the sky.

"What is that!" I say.

Erik, watching, says, "Even if I reported this no one would believe me."

Then we see the whale's tail, thin, narrow, far, far from the blow. It is in the air for a moment, then slowly it goes beneath and we realize we have seen a rare thing, a blue whale.

Like the grays, they are born tail first.

Long Sunlight, Many Whales

Kodiak Island, Summer Solstice 2000

Only after three years of following the gray whales up and down their migration on the West Coast did it truly come home to me—the *immensity* of their journey.

Look at how far the grays have traveled, I thought, as our plane soared over the glacial Chugach Mountains. Here in Alaska, the grays will fortify themselves, frolic, and feed, bathed in the seemingly eternal summer light of this Land of the Midnight Sun.

Linda and I had come to Kodiak Island, Alaska, to interview Eric Stirrup and biologist Stacy Studebaker who studies phytoplankton. Stacy had recently retired from teaching high school and had just rallied her students and the island community to bury a beached gray whale carcass for later re-articulation and display.

Many of us were concerned about the recent unprecedented die-off of migrating gray whales. The beached whales showed signs of starvation. We'll never know the cause of death for those who died at sea.

The whale deaths leave scientists with contending theories. Perhaps the deaths are caused by food scarcity, as the whales are drawn farther away from traditional feeding grounds because of resumed hunting and coastal development. Another theory comes from the National Oceanic and Atmospheric Administration (NOAA): Perhaps gray whale populations have rebounded so much the food supply has reached its carrying capacity. This means that the oceans may not support growing numbers of grays. Yet this theory does not take into consideration the huge pre-hunting populations of grays, since there are few sources except whale-hunting ships that can supply that valuable data. And there are other theories for the die-off as well—it may stem from pollution and toxic dumping. The gray whales may simply be giving us an early warning of the degenerating health of our oceans.

That's why we were here in these plentiful waters, where nine gray whales had washed up around Kodiak Island this spring. These mysterious whale deaths were the talk of this small town.

Kodiak Island, bordered by the Shelikof Strait, Kenai Peninsula, and the Gulf of Alaska, is often called the "Northwesternmost Hawaiian Island," or "The Emerald Isle." The 3,000 Kodiak brown bears living here are named after the island, which seems indeed an almost tropical paradise, a verdant jewel fortuitously set into the frigid gulf.

Everywhere in this island town is the influence of Native cultures—the Koniag, Aleut, and Alutiiq peoples have lived here for over 7,500 years. An impressive Alutiiq museum in the center of town is Native-run, as are many of the cultural heritage centers in Alaska; it offers insights into the complex cultures that existed before and still thrive after European settlement.

Very early on the summer solstice, we trooped down to the Kodiak harbor, which was crammed with boats of all kind—trawlers, gill-netters, long-liners, crabbers, and purse seiners. We also saw a multitude of sturdy charter boats with such names as *Pretty Woman II* and *Salmon King* and *Down the Hatch*. The morning newspaper had been full of Nome's famous Polar Bear Swim during their Midnight Sun Festival. In that frontier town, hardy locals plunge into the frigid 48-degree Bering Sea just south of the Arctic Circle to show off characteristic Nome bravado and to celebrate the solstice.

"There you are!" our charter captain Eric Stirrup called out cheerfully as we clambered aboard his boat, *Ten Bears*. "Coffee and sweets in the cabin. Help yourself and wake up." From his pilot's cabin, he and his deck hand, Lance, set about getting ready for what would be an astonishing day on the water.

But we didn't know what wonders awaited us. All we knew that morning was that it was very cold here in June, that all week it had poured rain all over Alaska, and that we were bleary-eyed and very hungry.

As if on cue, Stacy Studebaker ran down the boat ramp and jumped aboard. Within moments, the dockside ropes were loosened and we were off. With her light windbreaker, her dark-blonde ponytail, and sturdy boat shoes, 53-year-old Stacy looked as excited and expectant as if she were off on her very first whale-watching encounter. And yet this much-respected and popular high school science teacher has led many whale watches of her own. Her enthusiasm was so delightful it woke us up. At the same time, the marine mists began to clear, revealing a strong, sun-swept summer solstice day.

As we cruised out of Kodiak Harbor and away from the clouds over Kodiak, we settled on the stern to get acquainted with Stacy. We kept our binoculars handy, in case we spotted a blustery breath.

In all our whale-watching over the last years, we had kept a hopeful eye out for White Beak, the calf we met in the Baja birthing lagoons. By now he must be a strapping 20- or 30-ton juvenile. So every whale watch would begin with the mantra, "Keep a watch out for White Beak!" And after dead gray whales washed up along the West Coast—many of them juveniles—we'd ask for physical details. During the long migrations of the last two years, we prayed that White Beak was still a survivor.

Stacy and Lance talked about whales they've had the privilege of meeting. Stacy said, "You should have seen the grays rolling in Pasagshak Bay a week ago. My husband, Mike, and I couldn't believe it! Several grays in twos and threes came so close into shore where the water is really shallow. They actually made eye contact with us as they lolled about in the surf. I think they were playing! These giant creatures came right up close to us—not afraid, even somehow engaging with us."

Everywhere we looked now we saw seabirds, from shearwaters, to cormorants, to tufted and horned puffins with their comic antics. As we neared Pasagshak Bay, we kept watch for the familiar, heart-shaped blows, sure signs we were again among grays.

"When my husband, Mike, and I were staying at a friend's cabin on Pasagshak Bay, we not only saw those gray whales playing in the shallows, but we also spotted a dead whale out in the surf—a big, black body just floating along. We paddled out in our kayaks to see it better. The gray whale was on its right side, head in the water, pectoral flipper and midsection out of the surf. We paddled alongside and felt so sad for that whale."

In their small kayaks alongside the 40-foot, floating gray whale, Stacy and her husband watched each dead roll of the gray. Stacy felt helpless. Then she had an idea. "I thought, what if we could actually bury this whale and after nature had stripped its flesh and just left the skeleton, our town could put that whale's skeleton back together as a teaching tool and then display it in Kodiak for the future," Stacy said.

In what would turn out to be a true cooperative effort, Stacy and 15 other townspeople secured the proper permits. They called Mike Anderson, who owns a construction company and equipment. Anderson never balked at the size of the science project.

Because the whale was almost perfectly intact, still fresh,

with no damage from predators or decay, it was a perfect specimen for their science project, Stacy said. It was front-page news in the Kodiak Daily Mirror when Anderson used his earth-moving excavator to drag the whale to its grave. "Buried but not forgotten, the gray whale beached May 28 on Pasagshak Beach has found a not-so-final resting place," the newspaper article declared. "The whale's journey at last comes to an end—for a few years at least." Stacy estimated the male gray whale at 40,000 pounds, with tail flukes nine feet across. Together the workers wrapped the massive whale's flippers in a porous tarp and duct tape to make sure that in five to ten years they could retrieve all the bones.

"Once in the pit, before the crane started burying, it just happened," Stacy told us as we rocked along on the waves. "Everybody simply stopped and stood still around the rim of the pit. It was an incredible moment when everyone said thanks and paid their respects to this whale. We said goodbye." This moment of reverence was not planned. "Everyone naturally felt that this was the right thing to do."

At last the *Ten Bears* entered Pasagshak Bay, where we expected to meet up with the playful gray whales Stacy and her husband had seen just last week.

We all crammed into the pilot's cabin.

"Whoa!" Captain Eric grumbled. "Where are all the grays?"

Stacy leaned forward, peering through the windshield. "They were just here a few days ago."

Even with our binoculars riveted, we could not see a single gray whale in Pasagshak Bay. This emptiness after a summer of gray whale activity off Kodiak reduced us all to silence. We watched as Stacy proudly pointed out the site for the burial. I began to actually consider the prospect of returning from Alaska without having seen even one gray whale.

Eric was the first to speak. "I'm in the visual commodity business—and Pasagshak Bay today is a big bust." Then he brightened and turned the pilot's wheel a hard right. "But let's just head out to open sea." We chugged out of Pasagshak Bay into a gulf so smooth and sunlit, our moods could not stay dark for long. "You know, I always tell folks that if they just stay ashore on Kodiak to see the bears, they've missed the best wildlife viewing of all. Show me the bear that can leap two times its body weight and do a somersault, then land on his belly. Now that's something only whale-watchers can see!"

As we cruised out into open ocean, Stacy told us that many of these whales are summering around Kodiak and might be non-migratory residents. Stacy and her biologist friend Susan Paine are intrigued by the prospect of a resident pod of gray whales. "It's curious that the four whales washed up on our Kodiak beaches were all male. I wonder if they may have been part of a 'bachelor pod,' like the ones you see with other marine mammals, in walrus, dolphin, and sea lion societies. Perhaps if this residency is really true, we will one day soon be referring to Kodiak grays as well as Kodiak bears," Stacy said. "Or, they could all disappear tomorrow."

If the 100 or so gray whales summering off Kodiak Island

are indeed a growing group of residents, Stacy and Eric would be among their fiercest and most loyal allies. In our journeys following the gray whales along the West Coast, we'd met many wonderful "Whale People" as we came to call them. But Eric and Stacy are two of my favorites. Their enthusiasm, their diligent whale-watching and advocacy in a state not known for its environmental ethics or activism is remarkable. The grays have found good friends in these two islanders and in the Kodiak volunteers.

Amateur naturalists and wildlife advocates play a vital role in assuring the health of any species. And in a time when wildlife study has become so often political and polarized, those who study the whales simply out of affection and loyalty are often the most clear-sighted and forward-looking of all researchers. It is this grassroots commitment on the part of naturalists and local volunteers that can make all the difference in saving other species.

Cruising farther out from Kodiak into the Gulf of Alaska and around Ugak Island, Stacy noted the extraordinarily beautiful weather—calm, glassy seas and long, languid sunlight even here in the often-forbidding Gulf of Alaska. We all felt sun-blessed. "In all my 20 years of living in Kodiak," Stacy smiled, "I've never once even seen this side of Ugak Island. This is like seeing the dark side of the moon. Before today, it's always been bad weather, winds, or just not the right timing."

As we scanned the sunlit horizon for whales, Stacy continued, "During the war the Aleutian Islands were actually taken over by the Japanese and occupied when the Japanese invaded Attu and Kiska Islands at the far end of the Aleutian chain." We were familiar with these far-flung islands from studying the Alaska map. The archipelagoes of the Aleutian Islands have now been declared part of the Alaska Maritime National Wildlife Refuge. As these islands fan out like long fingers into the Pacific, there are few passageways for the gray whales heading north to the Bering Sea. One such narrow waterway in the Aleutians east chain is at the southern tip of Unimak Island. The grays migrate north through Unimak Pass, past the remote Pribilof and St. Lawrence Islands, past Nome and the border of the U.S. and Russia, past what used to be the ancient Bering Land Bridge, and finally way on up to the Arctic Circle and the Chukchi Sea.

Talk turned to the recent mortality of grays along the West Coast, including Alaska.

"We have such short memories for environmental damage done in the past, especially by the military," Stacy said. "But during and right after World War II, the U.S. military did a lot of terrible, toxic dumping off Kodiak. The exact location of some of this dumping is charted on marine navigation charts; other locations are lost. We'll never know because it was dumped in such a hurry. In fact, over a dozen toxic-waste sites have been identified on the island. And these sites are systematically being cleaned up by the Army Corps of Engineers, designated as SuperFund sites." She paused. "The government is coming around to taking some responsibility for this. But half a century later, I think we are only now beginning to see the effects. Perhaps the grays feeding on the ocean bottom are showing us, with their deaths, the health of our marine environment."

Another problem that may be affecting the whales could be toxic algae blooms, Stacy noted. "They occur all over the world. Certain species of phytoplankton reproduce very quickly under certain conditions, carrying strong toxic chemicals. In high concentrations, these blooms cause fish kills and bird mortality. In California they've even affected sea lions."

With the increase of POPS (persistent organic pollutants, such as DDT, PCBs, dioxins, lead, and benzene) in the ocean ecosystem, marine mammals may have weakened immune systems and be more susceptible to diseases and other changes in their environment, Stacy added. "Another piece of the puzzle is the sound pollution from man-made sources, such as the Navy's LFA sonar. This could all be affecting marine mammal health. As far as the gray whale deaths, you can't rule out anything until there is more study. And the tragic part of this is that not enough people are studying this die-off. Nobody that I'm aware of in Alaska has done anything but take tissue samples."

Our world's oceans hold a frightening amount of radioactive wastes. The Arctic Ocean, particularly the Barents and Kara Seas portions, has been a dumping ground for Russia's nuclear navy since the 1950s. A recent case study on Arctic Ocean dumping states that "Today scientists are trying to assess what possible damage the dumping might have done to the fragile environment of the Arctic region."

Now the waters were so calm, the Gulf of Alaska seemed no more than a gigantic lake with little land in sight. As we motored now very far out into the Gulf, I felt grateful to our skipper and biologist as they navigated what must be uncomfortable waters for Alaskan environmentalists. Many Alaskans are hostile to the intrusion of what a wolf trapper called, "Lower 48 Eco-Terrorists." There is often a Last Frontier sense here, a feeling that Alaska should be for Alaskans and not a national park paradise for the rest of the country. The fact that much of Alaska is federally owned and under the protection of the U.S. Forest Service or the National Park Service can make for some tense times for Alaska's environmentalists.

"Look, over by Cape Chiniak." Eric suddenly pointed across the gently lapping waves to a distant blow, a breath we'd long waited to see. "The whales have come back!"

Sure enough, there were two familiar, visible breaths off our bow as we recognized the gentle knuckle ridges and slopes of long backs gliding just barely above the waves. Barnacles like living adornments festooned each silver snout as the grays floated nearer our boat. Then a dramatic tail fluke rose and slapped the water as one whale dove below. We all cheered in delight at meeting up with the grays so far out at sea.

But then the whales came even closer, several on each side of our boat as if to accompany us out into such deep waters. At a speed of 13 knots and over waters that Eric noted were 50 fathoms deep, we were not in our element. We were land mammals in a small motorboat already 12 miles out into waters known for their depth and unpredictability. But we felt safe with our guardian gray whales alongside. In perfect synch, one whale on each side of us surfaced and blasted a huge mist of breath over our boat.

We all gasped at the intense whale breath.

"Great smell, eh?" Eric teased.

"It's like a combination of strong sulphur dioxide and a rotten, mud-swampy smell," Stacy said, and then explained. "You're smelling the sea bottom in their breath—a deep ooze because plankton dies and settles to the bottom to make organic mulch. Imagine layers of ooze and the grays with their baleen shoveling along the ocean bottom. Fossil fuel begins down there in that ooze rich from silt, mud, and volcanic ash."

We'd been misted before by gray whale breath in the birthing lagoons of Baja. But there was no odor to the Baja whales' breath, except a fresh and salty scent. Nor did the newborns, gaining daily bulk and nurturance from nursing on their mother's rich milk, ever smell.

In Alaska I didn't mind the pungency of gray whale breathing. Not only were we overjoyed to see them at last in their far north waters, but we also knew from their smell that they were feasting. And to witness the gray whales feeding was rare indeed. What made this sighting exciting was that the whales were diving so deep, tail flukes everywhere; they would return to the surface with mouths full of mud and tiny shrimp from the bottom, squeezing them out through the baleen on the surface. We could see the bubbles and the clouds of mud all around them where they were surfacing.

"It's very unusual to actually see gray whales feeding at this depth of 40 to 50 fathoms out here on the continental shelf,"

Stacy explained to us. "Usually you witness gray whales feeding closer to shore in more shallow water. Before whaling started, this deep-water diving may have been normal gray whale behavior. And as their population returns to their pre-hunting numbers, maybe they are resuming their normal feeding patterns. So what may seem rare to us may just be our lack of long-term observation and limited experience with this species.

"We haven't seen so many starving grays up north as you have down along the West Coast," Stacy noted. She paused as we watched more whales come our way. "Maybe we'll know more about what's really happening with the gray whales by the time we dig up and put back together our own whale skeleton on Kodiak."

As we all sat on the bow, drenched in sunlight and whale breath, we were surprised by how close the whales stayed by our boat. "Maybe the whales know we're part of the Friendly People Syndrome," Eric suggested.

Another whale dived and a mighty tail fluke raised up to stroke the water with an audible slap. The sound reminded me of an echo of a glacier calving. Suddenly three whales surfaced off the starboard side, their snouts raised just enough up out of the water to make eye contact with these strange humans who hopped about on the bow like excited children.

One of the whales was smaller, a juvenile who kept a big and very round eye on us as we all cruised along together. Their intense scrutiny reminded me of the Friendlies in Baja. But we were thousands of miles away from those protected lagoons and far out into waters that for whales and people are very dangerous.

Yet these two very large grays and juvenile kept pace with us as if it were the most natural alliance on earth. The juvenile raised his very pale snout higher and higher to take us in. He seemed to actually be waiting for us to recognize him, as he had already recognized us. "Oh, my God," Linda said. "It's White Beak!"

With his almost albino snout and even without his baby whiskers, White Beak was at last meeting up with us again after two years. We jumped up and down on the deck, and called out to this whale we'd known as a calf. He had survived and even seemed quite content accompanied by two huge grays. White Beak was in very good company.

We told Stacy the story of White Beak and how we'd been searching for him everywhere along the migration path. To at last find him well and thriving in Alaskan waters seemed a miracle. Whereas in the shallow Baja lagoons, White Beak and his mother had been able to spy-hop by simply resting their tail flukes on the sea floor, eyes just above the surface to better study us, now the juvenile was able to perform not only a spy-hop over 50 fathoms of ocean, but also to propel himself up into a breach. White Beak dazzled us with his airborne acrobatics, falling back into the waves with a huge flop, water streaming from his baleen.

White Beak again surfaced between the two larger grays and in one thrilling movement, all three whales dove in synch. Three tail flukes rose up above our boat, poised midair like three tall sails, and then waved goodbye, all in harmony.

"Now look," Eric pointed with a huge grin, "at nine, ten, eleven, twelve, one, two, three o'clock—more whales than you ever dreamed of seeing. And we're right smack dab in the middle of it!"

In our excitement over meeting up again with White Beak, we had not paid much mind to the other whale activity. But Eric was right. Straight ahead, spaced close together on the horizon, were the heart-shaped blows of over 30 whales. In a watery parade swirling around us were more whales at once than any of us had ever seen in our lives.

Whales were everywhere—off our starboard rail and our bow, surrounding our bobbing boat in concentric circles of blows and flukes, and ranging as far out on the 360-degree horizon as we could see. There were even whales beneath us, feeding fathoms deep. We could not count them fast enough before more whales surfaced, exhaling in blasts of breath and mist, sucking in the inspiration of air, then diving again, tail flukes streaming.

"We've been going through tier after tier of about 30 or 50 whales at once since you spotted White Beak," Eric happily told us. "I've been trying to keep count, but you know, I'm just amazed. There must be over 150 or even 200 gray whales way out here on the continental shelf. No wonder there were none left at Pasagshak Bay. They all decided to come out here on this fine summer's day and sunbathe in the deep sea."

Later that week, aerial surveys checked our sighting. Their estimate was 400 whales. For now, I tried to keep my balance on our boat, rocked up and down more by the churning, rolling, and diving of whales than by waves. These stunning multitudes of gray whales were arrayed around us more like a vision than a sighting. I could barely fathom what I was witnessing. Only a few other times in my life have I had this lucid, but hallucinatory sense that what I was seeing I would never completely take in. It was more like a living dream, but this was vividly real: Sailing through layers of great, generous grays was like going back through time to an ancient ocean alive and teeming with whales. A time before human history had done its deep harm. A time even before humans had a history. A time and a whole world of ancient whales.

Floating amidst this watery realm of gray whales was like an initiation into the mysteries of our ancient cetacean kin. Who could explain it? Yes, this was that rarely documented event of gray whales feeding over deep ocean, something usually observed only by those few of us who make it to Antarctica. But for all of us on that small boat afloat in a mass gathering of whales, it was more. It was like seeing the world whole again, if only for one long day.

"It must have always looked this way in the deep ocean before the whales were hunted," Linda said. "In pre-whaling days."

"Snowflakes," I said and everyone understood. This is what the Inupiat must have seen centuries ago: dappled and bright gray whales with their barnacled snouts rising up so abundantly it looked exactly like a snowstorm at sea. A mighty storm of whales in a wide, pristine sea.

"I will never forget this," Stacy said quietly. "I'll never forget all these whales."

Perhaps, like White Beak, these whales will help us to remember the ancient bond between peoples of the sea and peoples of the land as the grays navigate between so many worlds, seen and unseen.

The Mystery of the Humans

On man: Thus it is that these giants among giants have fallen beneath his arms.... In vain do they flee before him; his art will transport him to the ends of the earth; they will find no sanctuary except in nothingness.

RICHARD ELLIS, MEN AND WHALES

In the desert sea, why is it the gray whales bring their babies to us as if to say, Let us show you our young, aren't they grand, aren't they dark and shining and worth living? How different we are as a species that in one place we are out in a skiff watching the rainbow in their blow, caring greatly about them, and in another place we meet them with the offering of pain and death. In the desert lagoon of their births, I smell their breath and see the nutrient-rich mud from the bottom streaming from the mouth of the mother whale and then she turns and rolls again into the seamless movement of water and I think we shouldn't be here, we shouldn't let them know the double nature of our species. Love may be just as dangerous to their world as hatred or greed. They should fear us all, each and every one of us, for a human is a mystery.

The Limits of the Eye

Charles Melville Scammon and the other whale hunters had a narrowness of life, a smallness of sight. We seek to understand how a human creature can be so small as to kill a being so large, and can believe, even though a mother tries to protect her infant, that it has no love, no pain, no sentience, and that he does no wrong in this act against creation.

Scammon was an example of the limits of the eye, the split self, how a world of mystery and beauty was also a world to be killed and unraveled in a frenzy of blood and pain.

And that this is not every man is also a mystery, even though a welcome one. We are a species able to laugh and love, intelligent and often humane. Yet looking back, we ask what kind of men can take the light out of the eye of a whale without care, with bloody precision, and no respect or reverence? They were sighting without seeing, through crosshairs and with the calculated throw of a harpoon, or lancing or shooting or bombing.

It is as if there are distances in people, not only distances from their homes or across oceans or lands, but from themselves, distances between the heart and the mind.

There are those who keep wet a beached whale and try to return it to the water, those hunters who valiantly cut through ice to free a path for the gray whales caught in the Bering Sea. There are those who risk their lives for the whales, a girl who puts herself in the way of a harpoon. And then there are those like the man who put a cigar in a whale's blowhole or another who carved his name into a living beached whale.

Few recognize the delicate balance of this world. Humans have searched for, looked for, any sighting, cities of gold, fountains of youth; all these failed. We want to know where we come from and if there is uncharted life, unseen places, in the world, other universes.

We are a restless species.

The Limits of the Eye 2

The mystery of the human has many facets. We love or hate without reason. We bedevil the world in all meanings of the word, in so many ways, as if possessing it with evil and wrongdoing from within our own selves, corrupting it with human corruptions then cutting it down to our size. We see ourselves in nature reflected, while underneath its movement is a vision of greatness and immensity we cannot understand. We make the ocean, also, one of our nature, a place loving or hating, a place angry or calm. Adverse. Benevolent. Its many moods invented and named by men and their double emotions as if their inward spaces were beneath the waters.

Older maps portray whales as monsters, some with scales, some with great teeth instead of baleen. The monsters are spewing water from their mouths, not their blowholes. They live in an unknown element, near Terra Incognita, unknown land. On the maps, they are large enough to open their mouths and swallow the greatest of galleons. This is not only a matter of perspective. We can only imagine the fear the Europeans must have carried within as they approached this continent.

For the voyagers' fears were gliding on the ocean which so encircled them. The had depths that remained a secret, but the mystery of the human is that he has to search it, not to love it, not to care for it, but to see what is there for the taking, intruding on a mystery. These are all stories of the restless, those without content, content, in both meanings of the word.

For many people, the whale carried the weight of their own evil, a fear and judgment not based in reality. This is revealed even in the words: Leviathan. Behemoth. Monster. Beast. Behind the animal lay not majesty. The animal was there only for the humans. I think of the English queen who had an elephant and a rhinoceros fight to relieve her of boredom; our entire human history with animals has been aslant.

Yet for others, knowing its ways and mysteries, the whale is a sacred being. It is sentient. There is no doubt. It is part of a great creation. The sacred, someone once said, is that which can be destroyed but not created. Certainly that has been true of our relationship with whales, one of a long destruction; they are sacred and unable to be again created.

On this continent animals had special lives and powers. All nations of indigenous peoples had intelligent relationships with the animals. This is so of the first people of all continents. Their knowledge came from centuries of understanding, observing, care-taking.

So little was known about this world to the Europeans when they arrived. Perhaps history might have unfolded differently if they had known there was an ethic in dealings with animals. However, as events did unfold, never a thought was given to animals, their pain, their qualities of being. The world was illuminated by men in the shadows of their own lives, lives without reflection. Ecology and harmony within a working system; these were late sciences for the new arrivals, yet they are concepts that, in our time, are becoming alive again.

Over my lifetime, even from childhood, I've asked myself daily what is a human being, and in this musing and wondering, found that we are motivated by stirrings not always based in love or care or humanity. Elder writer John Hay wrote that he once believed the world was ruled by rationality and logic; now he believes it is ruled by rage and fear. I suppose we can also add greed, desire, and pride. Invisible, perhaps not even conscious, these are such large emotions in so small a being as a man or woman. Perhaps there are opposing currents, even riptides, inside a human. And in the same being who has these currents, there is still sometimes empathy. We find an animal enchanting that at other times we called beast.

How did so many come to be solitary wanderers in this great world with all the other kinds and shapes of life? When was it we came to think the world around us didn't count? Yet with each loss, there is a corresponding loss within the human. We, too, are diminished by what is forever gone, emptiness a part of the human, earthly condition.

We First People, some of us without stolen minds, still see the world as alive.

~

I think of men turning over in bed on ships, dreaming, or the captain with a lantern of whale oil, the light in the corner, his clothes over the chair. The human sleeps. In the bed, on the floor, a cot, there is the smell of the human body. At night, the mind is anchored to nothing except memory, dream, healing and want.

Inside a ship, perhaps a tablecloth is white, at least in one room, while there is slaughter on the deck, men's boots sliding in the blood, intestines being thrown into the red water. Men sick with blood and maybe seeing mother's milk, sick with the fear of the last thrashing of a dying creation, watching the throes of pain, the sounds of the creature's tail hitting water. The captain perhaps is clean. There may be a foghorn or a light, clouds, or rain falling. Then they wash the deck and the whale as the flensers, those who cut the flesh, go to work. Perhaps one of these same men once held the fallen blue egg of a robin or put it gently back into its nest.

That so delicate a creature as a human being, so light, hairless, with aching feet, burdens shouldered, could carry such a force of dreams and desires and wanderings that it would change a world, and make it possible to destroy one, is almost unimaginable. Following only a heard rumor or false map, a human will yet travel the "ends" of the Earth, of space, of everything, searching, looking for that which can't be found, the largest squid, the darkest star. And yet we can't see the more than 5,000 lives in a spoon of ocean water. We search the far edges of the universe, under the sea, but fear the depths of the human self.

But now, born in swaddling, turned to armor and then back to swaddling at the end, the body is curved in bed as if to return to the water of beginnings. We, as humans, are burdened or lightened, hour by hour, act by act.

Epilogue: What We Will Become in the Future

I have no doubt that the First Nations along the coast had the whale as a constellation. The sky is always named by what is on Earth, or in the world of gods, giants, and true mythology

We shelter ourselves in whales as once, in the north, some built shelters of whale bone, the entrance an arc of rib from an animal both revered and needed.

On this continent, there were those who painted whales and other animals on rocks, carved them, connected with them in a ceremonial mystery. In whaling there were the women who were still, hidden, and silent, the villagers who spoke only gently as the abstinent, prepared men paddled toward the whale with songs and prayers, with spirits. This was when the world was still inspirited with love for the creatures.

There are stories of the creations of whales. They are the creatures touched by water. They are, according to science, something like our ancestors, and some think that with their hidden human hands and leg bones, the parts they no longer need, they are what we will become in the future.

Then at the end of a life, what used to be a man becomes lightness of air or nothing. Water closes in on what used to be a whale.

Green water, blue, gray-black, still or churning water. It moves but is always contained. In this way, you'd think a man and sea are alike, but they aren't. In the sea, as Cousteau said, everything is moral. Everything that is eaten or born is god. But it is not god when there is no exchange. Nature is always alive with trades, but humans have greatly failed in these exchanges—from Captain Cook, who ate all that the Native peoples gave him and grew angry when they stopped, to the present when we take without question from the sea, even for what we call science. And finally, for the humans, the last sighting, drawing conclusions with a pen, a map, a mind. Human eyes not able to see themselves, their own deeds. Maps. Charts. Mystery. What we do comes from the deep, unsearched shadows of selves. The French/Lithuanian poet Oscar Milosz wrote that "when he lacks sorrow and tears, man is dry, poor, and accursed." It must be that the hydra is a two-headed killer, and lives not in water but in air and walks on two feet. One head lacks sorrow, desires, and is accursed. The other marvels.

This is a world vast and beautiful. It is too often unloved. It is too often without a just portion of human awe. Out in the ocean, the

birds fly over, the shining nautilus rises at night from the deep bottom of the ocean unseen by us. It is lit by the moonlight. A manta ray lights the waters it travels by its own luminescence. Plankton moves in clouds of vertical migrations from the top of the water to its depths. A garden of eels stand upright on the sea floor. A whale breathes.

Acknowledgments

This is a book about connections between people and animals. There are many of all species to thank. Dr. John Lilly once wrote that his long and groundbreaking study of dolphins taught him much about being human. This, too, has been the gift to us from the whales we've followed.

The people who have helped in the journey of this book have worked tirelessly and have wholeheartedly lent us their stories and perspectives. We'd like to thank Julie Rubenstein whose vision first put us together with National Geographic, Editor-in-Chief Kevin Mulroy who kept steady the whole project and found us the perfect editor in Pat Daniels's understanding, delicate eye. We are very grateful to Melissa Farris for her elegant book design. Elizabeth Wales, Brenda's insightful agent, was this book's champion all along the way and Linda's long-time agent, Beth Vesel, helped us both navigate this complicated and surprising journey. We'd like to acknowledge the respected *Seattle Times* editor James Vesely, whose early support and publication of our double vision of the Makah whaling story helped inspire this book in two voices.

The photographers whose beautiful images bring the gray whale alive in these pages have generously let their work illuminate our words. Many thanks to L.A. Henderson, who traveled along the migration trail with us, to Phillip Colla, Bill Hess in the Arctic, Sue Flood of the BBC, IFAW's Jared Blumenthal, Errol E. Povah in British Columbia, Margaret and Chuck Owens of Peninsula Citizens for the Protection of the Gray Whale, Pendleton Powell Brown, Sandy Rosenberg, Leo Shaw, Katy Penland, President of the American Cetacean Society, Paul Radcliffe, and Alison Schulman-Janiger who diligently counts whales in the official Southern California census.

All along the way, many tribal peoples have been kind enough to invite us into their homes and their traditions, including Makah elders Alberta Thompson and Dottie Chamblin, the whaling crew member Micah McCarty and his father, John, who at one time directed the Makah Whaling Commission. We thank Quileute Tribal Chairman Russell Woodruff and his visionary brother Fred Woodruff, whose wife Jill, daughter Sharra, and son Dakwa have paddled the "waterways of the ancestors" and kindly shared one journey with us in their family canoe. And in British Columbia, the artist and our guide Qaamina of the Ahousaht First Nations found the one gray whale that day in Clayoquot Sound and shared with us his and his father's, Stanley Sam Sr., vision for a future that benefits both native peoples and gray whales.

We have also been instructed and helped by researchers, peer-readers, and editorial assistants, including Dr. Toni Frohoff, Dr. Naomi Rose of the Humane Society of the U.S., Will Anderson of PAWS, naturalist/writer and researcher extraordinaire Leigh Calvez, artist Christine Lamb, researchers Lesa Quayle, Tara Kolden, and Bill McHallfey, and the expert, editorial encouragement of Maureen Michealson of NewSage Press who has generously allowed sections of Brenda's essay collection *Singing to the Sound: Visions of Nature, Animals, and Spirit* to reappear here in a different version.

To sister writer Susan Biskeborn, who has been Brenda's long-time first reader for 15 years, we both give humble and heartfelt thanks. Editors Marilyn Auer of the Bloomsbury Review and Professor Scot Slovic, of the University of Nevada's Center for Environmental Arts and Humanities, both gave invaluable critiques and editorial guidance during an early draft of this book. And Marlene Blessing put aside her own writing and editing to lend us her inspired sightings on this book at a crucial, final stage.

Linda would like to praise Kris Russell for taking care of her Colorado mountain home, horses, dog, and cats while she traveled and Brenda thanks Eve Anthony for abiding sea-shelter Vanessa Adams for keeping her waterfront studio and animals safe from Northwest tsunamis. Linda would like to thank her brother, Larry, who shares with her the astute eye for animals of their Chickasaw ancestors. To her parents, Brenda is grateful to her father for his lifetime of work conserving wildlife and to her mother for always listening to her stories.

And finally, we'd like to thank White Beak, a gray whale calf who, with umbilical cord still attached, had the courage and curiosity to greet us like an ambassador from another culture. By placing his trust in us, this newborn also gave birth to this book.